AF625076

Multiple-Choice Questions in Control Systems Engineering

S. Palani

Multiple-Choice Questions in Control Systems Engineering

Ane Books Pvt. Ltd.

Springer

S. Palani
Department of Electronics
and Communication Engineering
Sudharsan Engineering College
New Delhi, Delhi, India

ISBN 978-3-031-93786-6 ISBN 978-3-031-93787-3 (eBook)
https://doi.org/10.1007/978-3-031-93787-3

Jointly published with Ane Books Pvt. Ltd.
The print edition is not for sale in Afghanistan, Bangladesh, Bhutan, China (Mainland), India, Myanmar, Nepal and Pakistan. Customers from Afghanistan, Bangladesh, Bhutan, China (Mainland), India, Myanmar, Nepal and Pakistan please order the print book from: Ane Books Pvt. Ltd.
ISBN of the Co-Publisher's edition: 978-81-19160-68-6

© The Editor(s) (if applicable) and The Author(s), under exclusive license to Springer Nature Switzerland AG 2026
Previously published by Ane Books Pvt. Ltd.

This work is subject to copyright. All rights are solely and exclusively licensed by the Publisher, whether the whole or part of the material is concerned, specifically the rights of reprinting, reuse of illustrations, recitation, broadcasting, reproduction on microfilms or in any other physical way, and transmission or information storage and retrieval, electronic adaptation, computer software, or by similar or dissimilar methodology now known or hereafter developed.
The use of general descriptive names, registered names, trademarks, service marks, etc. in this publication does not imply, even in the absence of a specific statement, that such names are exempt from the relevant protective laws and regulations and therefore free for general use.
The publishers, the authors, and the editors are safe to assume that the advice and information in this book are believed to be true and accurate at the date of publication. Neither the publishers nor the authors or the editors give a warranty, express or implied, with respect to the material contained herein or for any errors or omissions that may have been made. The publishers remain neutral with regard to jurisdictional claims in published maps and institutional affiliations.

This Springer imprint is published by the registered company Springer Nature Switzerland AG
The registered company address is: Gewerbestrasse 11, 6330 Cham, Switzerland

If disposing of this product, please recycle the paper.

Preface

In recent times admissions to various kinds of jobs and higher studies including doctoral degree are done by conducting written examinations in the form of Multiple Choice Questions. State and Central government public sector under takings and private sectors invariably adapt the above method for preliminary selections. Students belonging to EEE, ECE, EI, ICE, and Chemical Engineering are tested whether they have in depth knowledge in Control Systems Engineering.

To meet the above requirement I have written this book which will be a good guide to them. About 600 questions have been given with answers. I have taken at most care to see the appropriateness and correctness of the questions and answers. I take this opportunity to thank Mr. Sexana, ANE Books Pvt. Ltd., New Delhi and Shri. Rathinam, General Manager, ANE Books Pvt. Ltd., Chennai. I would also like to thank Mr. V. Ashok, who typed the manuscript. I would also like to thank my wife Dr. S. Manimegalai, M.B.B.S., M.D. (Retired Joint Director of Medical Services) for helping me in the proof reading.

New Delhi, India S. Palani

Contents

About the Author

Dr. S. Palani earned his Bachelor of Engineering (B.E.) degree in Electrical Engineering from the University of Madras in 1966. He then obtained a Master of Technology (M.Tech.) in Control Systems Engineering from the Indian Institute of Technology Kharagpur in 1968 and completed his Ph.D. in Control Systems Engineering at the University of Madras in 1982. With over four decades of teaching experience, he began his career in 1968 at the Regional Engineering College (now known as the National Institute of Technology) in Tiruchirappalli, where he was part of the Electrical and Electronics Engineering (EEE) department. During his tenure, he held various positions, including Professor and Head of the department, and he played a crucial role in establishing the Instrumentation and Control Engineering department. After more than 30 years of dedicated service at REC Tiruchirappalli, he took on the role of Founder Principal at Sudharsan Engineering College in Pudukkottai. There, he successfully established multiple departments and developed extensive infrastructure. He has published over a hundred research papers in reputed international journals. Under his guidance, 17 research scholars were awarded Ph.D.

Chapter 1
System Transfer Function

Multiple Choice Questions

Choose the Correct Answer

1. **A system is defined as**
   (a) inter connection of objects.
   (b) inter connection of attributes.
   (c) inter connection of objects with a define relationships between objects and attributes.
   (d) none of the above.

2. **Example of an object is**
   (a) voltage generated by the generator.
   (b) an inductor.
   (c) current through an inductor.
   (d) the charge across an capacitor.

3. **Example of an attribute is**
   (a) temperature in a boiler bank.
   (b) current through an inductor.
   (c) speed of an electric motor.
   (d) all the above.

© The Author(s), under exclusive license to Springer Nature Switzerland AG 2026
S. Palani, *Multiple-Choice Questions in Control Systems Engineering*,
https://doi.org/10.1007/978-3-031-93787-3_1

4. Example of a system is

(a) interconnection of a mass dash-pot and a spring.
(b) water level, pressure and temperature in a boiler.
(c) current through an inductor.
(d) water discharge from a water bank.

5. A servomechanism is

(a) a system whose output is pressure.
(b) a system whose output is temperature.
(c) a system whose output is voltage.
(d) a system whose output is mechanical position.

6. A regulator system is one

(a) which maintains the output at the prescribed set level.
(b) which maintains the output as the input is varied.
(c) whose output varies as time increases.
(d) whose output is not correlated with respect to time.

7. An open loop system is one whose

(a) output depends on the error.
(b) control action does not depends on output charges.
(c) output is independent of control action.
(d) output varies linearly as the input varies.

8. A closed loop system is one whose

(a) output varies according to the difference between the input and output.
(b) the control action or actuating signals depends upon the changes in the output.
(c) the output depends upon the actuating signal.
(d) all the above.

9. An open loop system is

(a) less stable compared to closed loop system.
(b) more stable compared to closed loop system.
(c) very fast acting.
(d) more accurate.

10. Which of the following statements are true?

(a) An open loop system is fast acting and relatively more stable.
(b) An open loop system is sluggish and relatively more stable.
(c) An open loop system is fast acting and relatively less stable.
(d) An open loop system is sluggish acting and relatively less stable.

11. A closed loop system is

(a) fast acting and relatively less stable.
(b) fast acting and relatively more stable.
(c) sluggish but more stable.
(d) sluggish but less stable.

12. Find the correct statement.

(a) If the open loop system is stable, then the closed loop system is also stable.
(b) If the open loop system is stable, then the closed loop system is also unstable.
(c) If the open loop system is unstable, then by proper feedback the closed loop system can be made stable.
(d) If the open loop system is unstable, then the closed loop system is also unstable.

13. Positive feedback is not used in closed loop system because

(a) it is expensive.
(b) it introduces unwanted error.
(c) it makes the system unstable.
(d) it is insensitive to variation of parameters.

14. The unique advantage of closed loop system over open loop system is

(a) by proper feedback the unstable open loop system can be made stable.
(b) the closed loop system is more sensitive to variation of parameters.
(c) environmental variations do not affect the closed loop system performance.
(d) the closed loop system is fast acting compared to sluggish response of open loop system.

15. The transfer function of a system is

(a) the ratio of output to output.
(b) the efficiency of the system.
(c) the overall gain of the system.
(d) the ratio of the Laplace transform of the output variable to the Laplace transform of the input variable with initial conditions being zero.

16. The transfer function is said to be strictly proper if

(a) it obeys superposition theorem.
(b) the order of the denominator polynomial is greater than the order of the numerator polynomial.
(c) the order of the numerator and denominator polynomial is equal.
(d) the order of the numerator polynomial is greater than the order of the denominator polynomial.

17. A transfer function is said to be proper if

(a) it obeys superposition theorem.
(b) the numerator and denominator polynomials have the same order.
(c) the order of the numerator polynomial is greater than the denominator polynomial.
(d) the order of the denominator polynomial is greater than the order of the numerator polynomial.

18. A transfer function is said to be improper if

(a) it does not obey superposition theorem.
(b) the order of the numerator polynomial is greater than the order of the denominator polynomial.
(c) the order of the denominator is greater than the order of the numerator polynomial.
(d) the numerator and denominator polynomials have same order.

19. A transfer is said to be of minimum phase if

(a) all the poles are zeros are placed in the left half of the s-plane (LHP).
(b) all the poles are zeros are placed in the RHP of the s-plane.
(c) all the poles are zeros are placed in the LHP of the s-plane.
(d) all the zeros are placed in the LHP of the s-plane.

20. A transfer function is said to be of non-minimum phase if

(a) at least one pole is placed in the LHP of the s-plane.
(b) at least one zero is placed in the LHP of the s-plane.
(c) at least one pole or a zero is placed in the RHP of the s-plane.
(d) at least one pole is placed in the LHP and a zero in RHP of the s-plane.

21. For an all pass transfer function,

(a) the poles and zeros are placed symmetrically with respect to the real axis.
(b) poles and zeros are placed symmetrically with respect to the $j\omega$ axis.
(c) poles and zeros are placed such that pole-zero cancellation takes place.
(d) repeated poles are placed at the origin of the s-plane.

22. Consider the following sinusoidal transfer function.

$$G(j\omega) = \frac{(j\omega + 10)}{(j\omega - 10)}$$

The magnitude and phase of the above transfer function is

(a) $G(j\omega) = 1 \underline{|\pi + 2\tan^{-1}\frac{\omega}{10}}$.

(b) $G(j\omega) = 1 \underline{|\pi - 2\tan^{-1}\frac{\omega}{10}}$.

(c) $G(j\omega) = 1 \underline{|\pi + 2\tan^{-1}\frac{10}{\omega}}$.

(d) $G(j\omega) = 1 \underline{|\pi}$.

23. Consider the following sinusoidal transfer function of an all pass system.

$$G(j\omega) = \frac{1(j\omega + 2)(j\omega + 3)}{(j\omega - 2)(j\omega + 3)}$$

The magnitude and phase of the transfer function is

(a) $G(j\omega) = 1 \underline{|\pi + 2\tan^{-1}\frac{\omega}{2} - 2\tan^{-1}\frac{\omega}{3}}$.

(b) $G(j\omega) = 1 \underline{|2(\tan^{-1}\frac{\omega}{2} - \tan^{-1}\frac{\omega}{2})}$.

(c) $G(j\omega) = 1 \underline{|2(\pi + \tan^{-1}\frac{\omega}{2} - \tan^{-1}\frac{\omega}{2})}$.

(d) $G(j\omega) = 1 \underline{|\pi}$.

24. The zero of a transfer function is

(a) the numerator polynomial of the transfer function.
(b) the denominator polynomial of the transfer function.
(c) the points in the s-plane at which the transfer function becomes zero.
(d) the points in the s-plane at which the transfer function becomes infinity.

25. The poles of a transfer function is

(a) the denominator polynomial of the transfer function.
(b) the numerator polynomial of the transfer function.
(c) the points in the s-plane at which the transfer function becomes zero.
(d) the points in the s-plane at which the transfer function becomes infinity.

26. In force-voltage analogy the mass is represented by the electrical element

(a) capacitor.
(b) inductor.
(c) resistor.
(d) voltage transformer.

27. In force-voltage analogy a mechanical spring is represented by the electrical element

(a) reciprocal of capacitor.
(b) inductor.
(c) resistor.
(d) voltage transformer.

28. In voltage-force analogy a dash-pot is represented by the electrical element

(a) capacitor.
(b) inductor.
(c) resistor.
(d) voltage transformer.

29. In voltage-force analogy the mechanical lever arrangement is represented by the electrical element

(a) capacitor.
(b) inductor.
(c) resistor.
(d) voltage transformer.

30. In torque-current analogy, the mechanical inertia is represented by the following electrical element

(a) inductor.
(b) capacitor.
(c) resistor.
(d) current transformer.

31. In torque-current analogy, the mechanical element spring is represented by the electrical element

(a) capacitor.
(b) reciprocal of inductor.
(c) inductor.
(d) current transformer.

32. In torque-current analogy, the rotational dash-pot is represented by the electrical element

(a) inductor.
(b) capacitor.
(c) reciprocal of resistance.
(d) current transformer.

33. In torque-current analogy, mechanical gear arrangement is represented by the electrical element

(a) inductor.
(b) capacitor.
(c) resistor.
(d) current transformer.

34. The advantage of using armature controlled d.c. motor in server application is

(a) it has fast transient response and provides additional damping.
(b) the efficiency is very high.
(c) it does not require any amplifier in the armature circuit.
(d) the cost is low.

35. The disadvantage of field controlled d.c. motor is

(a) it has poor efficiency.
(b) the cost is high.
(c) time response is sluggish due high field time constant.
(d) the size of the motor is large.

36. The advantage of using two phase servo-motor is

(a) its efficiency is high.
(b) the negative torque-speed characteristic provides positive damping and hence ensures stability.
(c) the cost is low.
(d) small size.

37. High resistance is intentionally introduced in the rotor ride of a two phase servo motor to

(a) dissipate the excessive heat.
(b) improve the efficiency.
(c) make the slope of the torque speed characteristic negative for stable operation.
(d) compensate for voltage drop in the rotor circuit.

38. For servo applications, servo motors are used in preference to D.C. motors because

(a) the efficiency is high.
(b) mechanical time constant is low.
(c) the cost is low.
(d) electrical time constant is high.

39. The servo motor is designed to have negative torque-speed characteristic because

(a) it ensures low cost.
(b) it time constant is low.
(c) it ensures stable operation at all speeds.
(d) its torque is increased.

40. The rotors of synchro generator (SG) and control transformer (CT) have

(a) both dump bell shape.
(b) SG has dump bell and CT has cylindrical shape.
(c) both have cylindrical shape.
(d) SG rotor is cylindrical and CT dump bell shaped rotors.

41. The stators of SG and CT are wound by three phase connections. The voltage induced in the CT rotor is

(a) single phase.
(b) two phase.
(c) three phase.
(d) rectangular wave.

42. With respect to the angular difference between the SG rotor and CT rotor the output voltage across CT rotor is

(a) step voltage.
(b) ramp voltage.
(c) sinusoidal voltage.
(d) rectangular wave.

43. The CT rotor is specifically designed to be cylindrical shaped to get

(a) constant output voltage.
(b) the change in rotor impedance with the rotor position is minimized.
(c) the air gap flux unevenly distributed around the rotor.
(d) maximum power transfer.

44. A pair of synchros is mainly used as

(a) voltage changing device.
(b) error detector.
(c) current changing device.
(d) position measuring device.

45. In a torque-voltage analogy, the angular velocity is analogous to

(a) loop voltage.
(b) loop current.
(c) nodal current.
(d) nodal voltage.

46. In a force-current analogy the translational velocity is analogous to

(a) loop current.
(b) loop voltage.
(c) nodal current.
(d) nodal voltage.

47. The main disadvantage of a stepper motor is

(a) it is a digitally controlled device.
(b) poor precision positioning.
(c) for high power applications it is not used.
(d) its efficiency is poor.

48. A certain stepper motor has a stator teeth 80 and rotor tooth 100. The angular displacement is

(a) 1.8°.
(b) 3.6°.
(c) 0.9°.
(d) 0.45°.

49. A two phase step motor has a rotor teeth of 24. Then the step movement is

(a) 2.5°.
(b) 5.0°.
(c) 7.5°.
(d) 10°.

50. A two phase stepper motor has 24 rotor teeth. The number of steps per revolution is

(a) 96.
(b) 24.
(c) 16.
(d) 48.

51. A stepper motor has a step angle of 1.8°. The speed of the stepper motor for 300 pulses is

(a) 120 rpm.
(b) 60 rpm.
(c) 90 rpm.
(d) 450 rpm.

52. The block diagram is a

(a) mathematical representation.
(b) photographic representation.
(c) pictorial representation.
(d) graphical representation.

53. In a block diagram, a block represents

(a) functional relationship between input and output.
(b) the ratio of output to input.
(c) efficiency.
(d) Laplace transform.

54. The main advantage of block diagram is

(a) it is easy to find the efficiency since the output and input of each block is known.
(b) power flow is easily identified.
(c) the functional operation is easily visualized without examining physical system.
(d) important functions are concealed within the walls of the block diagram.

55. The main disadvantage of block diagram represents is

(a) block diagram representation for a particular system is not unique.
(b) the main source of energy flow in the system is not shown in the block diagram.
(c) system where the input-output are not inter-changeable are not considered.
(d) system with nonzero initial conditions are not considered in block diagram.

56. In the signal flow graph (SFG), using Mason's gain formula one can find the total gain between

(a) output node and input node.
(b) output node and any other node.
(c) between any two nodes.
(d) all the above.

57. In Mason's gain formula when applied to SFG, the determinant Δ is defined for

(a) output to input variable only.
(b) single input, single output only.
(c) multi-input, multi-output only.
(d) Δ is same for all the above cases.

58. A non-touching loop in a SFG is defined

(a) as the path which originates from a node and ends at the same node without encountering the same node.
(b) as the path starting from a node and ends at the same node.
(c) as the loops which are not touching each other and do not have any common node.
(d) as any closed path.

59. There are five forward paths in a SFG. There are three non-touching loops. Δ_m and Δ for this SFG is

(a) $\Delta_m = 5$ and $\Delta = 8$.
(b) $\Delta_m = 5$ and $\Delta = 2$.
(c) $\Delta_m = 5$ and $\Delta = 1$.
(d) $\Delta_m = 8$ and $\Delta = 1$.

60. The type of a system is identified if the T.F. contains

(a) pure differentiator.
(b) pure integrator.
(c) there is at least a pole in LHP.
(d) there is at least a pole in RHP.

61. The order of the system is decided by

(a) the highest degree of the numerator polynomial of the T.F. of the system.
(b) the highest degree of the denominator polynomial of the T.F.
(c) the difference between the degree of the numerator and denominator polynomials of the T.F.
(d) the polynomial of the numerator or the denominator which ever has the highest degree.

62. A monic polynomial is a polynomial,

(a) whose free term in the characteristic equation is zero.
(b) whose coefficient of the highest degree of the characteristic equation is unity.
(c) all the coefficients of the characteristic equation are non-zero.
(d) at least one coefficient of the characteristic equation is negative.

63. The characteristic equation is obtained by equating the

(a) numerator polynomial of the T.F. to zero.
(b) denominator polynomial of the T.F. to zero.
(c) the sum of the degree of numerator and denominator polynomials of the T.F. to zero.
(d) the difference of the degree of numerator and denominator polynomials of the T.F. to zero.

64. What is an actuating signal?

(a) It actuates the power device directly or indirectly.
(b) The difference between the input and the output signals.
(c) The input to the amplifier used in the system.
(d) It is an error signal.

65. The advantage of open loop system is

(a) it is simple to design.

(b) components requirement is less.
(c) physically realizable open loop systems are generally stable.
(d) all the above.

66. The disadvantage of open loop system is

(a) it is manually operated.
(b) it is sluggish in operation.
(c) the accuracy is less.
(d) all the above.

67. The advantage of closed loop system is

(a) it is fast acting.
(b) static and dynamic errors are less.
(c) unstable open loop system can be made stable by appropriate feedback.
(d) all the above.

68. The disadvantage of closed loop system is

(a) more components connected in the system make it complex.
(b) it is costly.
(c) in some systems excessive over-shoots are observed before coming to steady-state.
(d) all the above.

69. What is an input?

(a) The signal which controls the power device.
(b) The attribute of the system which can be independently varied.
(c) It is the actuating signal in the system.
(d) The control action is done by the input signal.

70. What is the output signal?

(a) It is an attribute which cannot be varied independently.
(b) It is an attribute which can be varied by varying the input attribute.
(c) It is an attribute in the power device in the system.
(d) It is an output of the amplifier used in the system.

71. Give an example of a servo-mechanism.

(a) Control of pressure in a steam boiler.
(b) Control of fission reaction in a nuclear reactor.
(c) Positioning of an anti-aircraft gun.
(d) The temperature control in a chemical process.

72. Consider the following T.F.

$$G(s) = \frac{10(s^2 + 2s + 4)}{(s^3 + 5s^2 + 7s + 8)}$$

What type of T.F. it is?

(a) It is a proper T.F.
(b) It is strictly proper T.F.
(c) It is non proper T.F.
(d) It is all pass T.F.

73. Consider the following T.F.

$$G(s) = \frac{10(s^2 + 2s + 4)}{(5s^2 + 7s + 4)}$$

What type of T.F. it is?

(a) It is a proper T.F.
(b) It is strictly proper T.F.
(c) It is an improper T.F.
(d) It is an all pass T.F.

74. Consider the following T.F.

$$G(s) = \frac{10(s^2 + 2s + 4)}{(2s + 5)}$$

What type of T.F. it is?

(a) It is a proper T.F.
(b) It is strictly proper T.F.
(c) It is an improper T.F.
(d) It is an all pass T.F.

75. Consider the following T.F.

$$G(s) = \frac{K(s + 10)}{(s - 10)}$$

What type of T.F. it is?

(a) Strictly proper T.F.
(b) Improper T.F.
(c) All pass T.F.
(d) Minimum phase T.F.

76. Consider the following T.F.

$$G(s) = \frac{10(s + 2)}{(s + 3)(s + 9)}$$

What type of T.F. it is?

(a) It is a proper T.F.
(b) It is minimum phase T.F.
(c) It is improper T.F.
(d) It is all pass T.F.

77. Consider the following T.F.

$$G(s) = \frac{10(s+2)}{(s-3)(s+9)}$$

What type of T.F. it is?

(a) It is proper T.F.
(b) It is improper T.F.
(c) It is all pass T.F.
(d) It is non-minimum phase T.F.

78. The magnitude of all pass T.F. is

(a) $|G(j\omega)| = 1.$
(b) $|G(j\omega)| = 0.8.$
(c) $|G(j\omega)| = 0.5.$
(d) $|G(j\omega)| = 0.1.$

79. The forward path T.F. of a certain system is $G(s)$ and the negative feedback path T.F. is $H(s)$. The closed loop T.F. is

(a) $\frac{C(s)}{R(s)} = \frac{G(s)}{[1 - GH(s)]}.$
(b) $\frac{C(s)}{R(s)} = \frac{G(s)}{[1 + GH(s)]}.$
(c) $\frac{C(s)}{R(s)} = \frac{GH(s)}{[1 - GH(s)]}.$
(d) $\frac{C(s)}{R(s)} = \frac{GH(s)}{[1 + GH(s)]}.$

80. A certain non-unity, negative feedback system has forward path T.F. $G(s)$ and feedback path T.F. $H(s)$. This may be converted as a unity feedback system with the following forward path T.F.

(a) $Geq^{(s)} = \frac{G(s)}{[1 + GH(s) + G(s)]}.$

(b) $Geq^{(s)} = \frac{GH(s)}{[1 + GH(s) + G(s)]}.$

(c) $Geq^{(s)} = \dfrac{G(s)}{[1 + GH(s) - G(s)]}$.

(d) $Geq^{(s)} = \dfrac{GH(s)}{[1 + GH(s) - G(s)]}$.

81. If $T(s)$ is the closed loop T.F. of a unity negative feedback system, then the open loop T.F. is given by

(a) $G(s) = \dfrac{T(s)}{(1 + T(s))}$.

(b) $G(s) = \dfrac{T(s)}{(1 - T(s))}$.

(c) $G(s) = \dfrac{(1 + T(s))}{T(s)}$.

(d) $G(s) = \dfrac{(1 - T(s))}{T(s)}$.

82. The error function of a non-unity transfer function is given by

(a) $E(s) = \dfrac{R(s)[1 + GH(s) + G(s)]}{[1 + GH(s)]}$.

(b) $E(s) = \dfrac{R(s)[1 - GH(s) + G(s)]}{[1 + GH(s)]}$.

(c) $E(s) = \dfrac{R(s)[1 + GH(s) - G(s)]}{[1 - GH(s)]}$.

(d) $E(s) = \dfrac{R(s)[1 + GH(s) - G(s)]}{[1 + GH(s)]}$.

83. For a certain unity negative feedback system the closed loop T.F. is $T(s)$. The error function is given by

(a) $E(s) = [1 - T(s)]$.

(b) $E(s) = \dfrac{[R(s)T(s)]}{[1 + T(s)]}$.

(c) $E(s) = R(s)[1 - T(s)]$.

(d) $E(s) = \dfrac{[R(s)T(s)]}{[1 - T(s)]}$.

84. In a block diagram representation the system is subjected to input $R(s)$ and disturbance $D(s)$. In each case the characteristic equation will be

(a) same.
(b) totally different.
(c) it depends upon the location of the disturbance.
(d) it depends upon the location of the input.

85. Mason's gain formula in SFG is

(a) $A = \sum_{m=1}^{n} T_m \Delta_m$.
(b) $A = \sum_{m=1}^{n} \frac{T_m \Delta_m}{\Delta}$.
(c) $A = \frac{T_n \Delta_n}{\Delta}$.
(d) $A = \frac{\Delta}{T_n \Delta_n}$.

86. For a non-unity negative feedback system, Δ in Mason's gain formula is given by

(a) $\Delta = G(s)H(s)$.
(b) $\Delta = 1 + GH(s)$.
(c) $\Delta = 1 - GH(s)$.
(d) $\Delta = \frac{1}{1 + GH(s)}$.

87. For a non-unity negative feedback control system in Mason's gain formula,

(a) $T_m = 1$.
(b) $T_m = 1 + GH(s)$.
(c) $T_m = 1 - GH(s)$.
(d) $T_m = G(s)$.

88. For a non-unity negative feedback system, in Mason's gain formula,

(a) $\Delta_m = 1$.
(b) $\Delta_m = 1 + GH(s)$.
(c) $\Delta_m = 1 - GH(s)$.
(d) $\Delta_m = GH(s)$.

89. The nodes in SFG represent

(a) the parameters.
(b) variables.
(c) the inputs.
(d) the outputs.

90. In SFG the source

(a) has only outgoing branches.
(b) has only incoming branches.
(c) has both incoming and outgoing branches.
(d) none of the above.

91. In SFG the sink

(a) has outgoing branches only.
(b) has incoming branches only.

(c) has both incoming and outgoing branches.
(d) none of the above.

92. In SFG, the branch represents

(a) the system variables.
(b) the system parameters.
(c) the transmission of signal from one node to the other node.
(d) the input and output.

93. In SFG, the mixed node represents

(a) the input variable.
(b) the output variable.
(c) both input and output variables.
(d) nodes which do not represent either input node or output node.

94. In a signal flow graph the path

(a) is a collection of branches.
(b) is a connection of branches between input node and output node.
(c) is a continuous unidirectional succession of branches along which no node is passed more than once.
(d) is a continuous succession of branches.

95. In a mechanical system the independent variable $f(t)$ is analogous to the loop system in electrical network as

(a) voltage source $e(t)$.
(b) current source $i(t)$.
(c) voltage $e(t)$ and current source $i(t)$.
(d) nodal voltage $e(t)$.

96. In a mechanical system the independent force $f(t)$ is analogous to the nodal electrical network as

(a) the voltage source $e(t)$.
(b) the current source $i(t)$.
(c) voltage and current source.
(d) the loop current $i(t)$.

97. In a force-voltage analogy the lever arrangement with l_1 and l_2 lengths from the pivot is represented in electrical analogous network as a

(a) voltage transformer of turns ratio $l_2 : l_1$.
(b) voltage transformer of turns ratio $l_1 : l_2$.
(c) current transformer of turns ratio $l_1 : l_2$.
(d) current transformer of turns ratio $l_2 : l_1$.

98. In a force-current analogy the lever arrangement from l_1 and l_2 lengths from the pivot is represented in electrical analogous network as a

(a) voltage transformer of turns ratio $l_1 : l_2$.
(b) voltage transformer of turns ratio $l_2 : l_1$.
(c) current transformer of turns ratio $l_1 : l_2$.
(d) current transformer of turns ratio $l_2 : l_1$.

99. In a rotational mechanical system the gear arrangement has n_1 and n_2 teeth. In the torque-voltage analogy it is represented in the electrical network as a

(a) voltage transformer with a turns ratio of $n_1 : n_2$.
(b) voltage transformer with a turns ratio of $n_2 : n_1$.
(c) current transformer with a turns ratio of $n_1 : n_2$.
(d) current transformer with a turns ratio of $n_2 : n_1$.

100. In a rotational mechanical system the gear arrangement has n_1 and n_2 teeth. In the torque-current analogy it is represented in the electrical network as a

(a) voltage transformer with a turns ratio of $n_1 : n_2$.
(b) voltage transformer with a turns ratio of $n_2 : n_1$.
(c) current transformer with a turns ratio of $n_1 : n_2$.
(d) current transformer with a turns ratio of $n_2 : n_1$.

Answers

(1) c	(15) d	(29) d	(43) b	(57) d
(2) b	(16) d	(30) b	(44) b	(58) c
(3) d	(17) b	(31) b	(45) b	(59) c
(4) a	(18) b	(32) b	(46) d	(60) b
(5) d	(19) a	(33) d	(47) c	(61) d
(6) a	(20) c	(34) a	(48) c	(62) b
(7) b	(21) b	(35) c	(49) c	(63) b
(8) b	(22) a	(36) b	(50) d	(64) a
(9) b	(23) b	(37) c	(51) c	(65) d
(10) b	(24) c	(38) b	(52) c	(66) d
(11) a	(25) d	(39) c	(53) a	(67) d
(12) c	(26) b	(40) b	(54) c	(68) d
(13) c	(27) a	(41) a	(55) b	(69) b
(14) a	(28) c	(42) c	(56) d	(70) b

(71) c
(72) b
(73) a
(74) c
(75) c
(76) b
(77) d
(78) a
(79) b
(80) c
(81) b
(82) d
(83) c
(84) a
(85) b
(86) b
(87) d
(88) a
(89) b
(90) a
(91) b
(92) c
(93) d
(94) c
(95) a
(96) b
(97) a
(98) c
(99) b
(100) c

Chapter 2
Time Response Analysis

Multiple Choice Questions

Choose the Correct Answer

1. **Time response of a system is the**
   (a) transient response of the system as time t varies.
   (b) steady state response of the system as time t varies.
   (c) combination of transient response and steady state response as time t varies.
   (d) variation of only the output attribute as time t varies.

2. **The transient response of the system is**
   (a) the time response of the system which goes from initial stage to final stage with respect to time t.
   (b) the time response of the system at time t tending to infinity.
   (c) the variation of only the output varies as time t increases.
   (d) the response of the system which stop varying as time t varies.

3. **The steady state response is**
   (a) variations of all attributes as time t varies.
   (b) variations of only the output as time t varies to infinity.
   (c) variations of all attributes in the system as time t is infinity.
   (d) variations of only the output attributes as time t varies.

4. **The Laplace transform of an impulse function is**
   (a) $R(s) = 1$.
   (b) $R(s) = \frac{1}{s}$.

© The Author(s), under exclusive license to Springer Nature Switzerland AG 2026

S. Palani, *Multiple-Choice Questions in Control Systems Engineering*,
https://doi.org/10.1007/978-3-031-93787-3_2

(c) $R(s) = \frac{1}{s^2}$.
(d) $R(s) = 0$.

5. The systems are tested with an impulse because

(a) it gives good time response.
(b) impulse signal is easy to generate.
(c) if impulse response is satisfactory, it gives satisfactory response to other types of inputs.
(d) if the area under the impulse response is finite then the system is bounded input-bounded output (BIBO) stable.

6. The systems are tested for step input because

(a) it is easy to generate.
(b) if it gives good time response, the system will give satisfactory response for other inputs such as impulse and velocity.
(c) from step response, time responses due to other inputs such as impulse, ramp and acceleration can be easily obtained.
(d) all the above.

7. Find the correct statement.

(a) Impulse $\xrightarrow{\text{integrate}}$ Step $\xrightarrow{\text{integrate}}$ Ramp.
(b) Impulse $\xrightarrow{\text{differentiate}}$ Step $\xrightarrow{\text{differentiate}}$ Ramp.
(c) Ramp $\xrightarrow{\text{integrate}}$ Impulse $\xrightarrow{\text{integrate}}$ Step.
(d) Ramp $\xrightarrow{\text{differentiate}}$ Impulse $\xrightarrow{\text{differentiate}}$ Step.

8. Find the correct statement.

(a) Ramp $\xrightarrow{\text{integrate}}$ Step $\xrightarrow{\text{integrate}}$ Impulse.
(b) Ramp $\xrightarrow{\text{differentiate}}$ Step $\xrightarrow{\text{differentiate}}$ Impulse.
(c) Impulse $\xrightarrow{\text{differentiate}}$ Step $\xrightarrow{\text{differentiate}}$ Ramp.
(d) Step $\xrightarrow{\text{integrate}}$ Ramp $\xrightarrow{\text{integrate}}$ Impulse.

9. For what type of input, the time domain specifications are derived for first and second order system?

(a) Impulse input.
(b) Step input.
(c) Ramp input.
(d) Acceleration input.

10. A first order system is subjected to unit step input. The percentage overshoot is

(a) 100%.
(b) 50%.
(c) 25%.
(d) zero.

11. For a first order system, the time constant of the system has the following effect.

(a) For small time constant, the time response is sluggish.
(b) For high time constant, the time response is faster.
(c) For small time constant, the system time response is faster.
(d) For zero time constant, the system does not respond.

12. For a first order system with a time constant T, for step input, the rise time is given by

(a) $t_r = 0.693T$.
(b) $t_r = 2.2T$.
(c) $t_r = 3.91T$.
(d) $t_r = 2.66T$.

13. For a first order system with a time constant T, for step input, the time delay is given by

(a) $t_d = 0.693T$.
(b) $t_d = 2.2T$.
(c) $t_d = 3.91T$.
(d) $t_d = 2.66T$.

14. For a first order system with a time constant T, the settling time for 5% error tolerance is given by

(a) $t_s = 3.91T$.
(b) $t_s = 2.66T$.
(c) $t_s = 3T$.
(d) $t_s = 4T$.

15. Consider the following T.F. of a certain first order system

$$G(s) = \frac{10}{(s + 10)}$$

The time constant of the system is given by

(a) $T = 0.1$ sec.
(b) $T = 1.0$ sec.

(c) $T = 5.1$ sec.
(d) $T = 10.1$ sec.

16. A glass bulb thermometer reads 98.2° of the final value of temperature one minute after immersing in hot water. The settling time for 5% error tolerance is given by

(a) $t_s = 15$ sec.
(b) $t_s = 33$ sec.
(c) $t_s = 45$ sec.
(d) $t_s = 60$ sec.

17. For an under damped standard second order system, the two roots are placed at

(a) $s_{1,2} = \zeta\omega_n \pm j\omega n\sqrt{1-\zeta^2}$.
(b) $s_{1;2} = -\zeta\omega_n \pm j\omega n\sqrt{1-\zeta^2}$.
(c) $s_{1,2} = -\zeta\omega_n \pm j\omega n\sqrt{\zeta^2-1}$.
(d) $s_{12} = \zeta\omega_n \pm j\omega n\sqrt{\zeta^2-1}$.

18. For an over damped standard second order system, the roots are placed at

(a) $s_{1,2} = \zeta\omega_n \pm j\omega n\sqrt{1-\zeta^2}$.
(b) $s_{12} = -\zeta\omega_n \pm j\omega n\sqrt{1-\zeta^2}$.
(c) $s_{12} = -\zeta\omega_n \pm j\omega n\sqrt{\zeta^2-1}$.
(d) $s_{12} = \zeta\omega_n \pm j\omega n\sqrt{\zeta^2-1}$.

19. For a critically damped second order system, the roots are placed at

(a) $s_{1,2} = -\zeta\omega_n \pm j\omega n\sqrt{1-\zeta^2}$.
(b) $s_{1,2} = -\zeta\omega_n \pm j\omega n\sqrt{\zeta^2-1}$.
(c) $s_{12} = -\omega_n$.
(d) $s_{12} = \pm j\omega n$.

20. For an undamped second order system the roots of the standard second order system are placed at

(a) $s_{1,2} = \pm j\omega_n$.
(b) $s_{1,2} = -\omega_n$.
(c) $s_{1,2} = -\omega_n^2$.
(d) $s_{1,2} = \omega n^2$.

21. The presence of repeated poles in the $j\omega$ axis will result in

(a) sustained oscillations in the system.
(b) instability of the system.
(c) decaying oscillations without loss of stability.
(d) none of the above.

22. What is an insignificant pole?

(a) A pole placed in the negative real axis and closer to the origin in the complex s-plane.
(b) Poles placed in the $j\omega$ axis.
(c) Poles placed far away from the $j\omega$ axis.
(d) Repeated poles in the $j\omega$ axis.

23. What are dominant poles?

(a) Poles placed at the origin of the s-plane.
(b) Poles placed in the negative real axis.
(c) Repeated poles in the $j\omega$ axis.
(d) Complex conjugate poles placed in the left half of the s-plane (LHP) nearer to the $j\omega$ axis are called dominant poles. All other poles have real part to the left of these poles.

24. What is pole-zero cancellation?

(a) If a pole moves closer to the zero pole zero cancellation takes place.
(b) If a zero moves towards the pole, then pole zero cancellation takes place.
(c) If both pole and zero move far off from the $j\omega$ axis, pole zero cancellation takes place.
(d) If the location of a pole and a zero are same pole zero cancellation takes place.

25. The effects of pole-zero cancellation are

(a) the internal stability is lost, system becomes uncontrollable and unobservable and the order of the system is reduced.
(b) the order of the system is reduced.
(c) the system becomes uncontrollable and unobservable.
(d) the internal stability is lost but external stability retained.

26. The imaginary part of a pole generates

(a) sinusoidal response.
(b) impulse response.
(c) step response.
(d) exponential decay or rise.

27. In a system, the real part of pole generates

(a) sinusoidal response.
(b) exponential decay or rise.
(c) impulse response.
(d) step response.

28. For a second order prototype system with constant imaginary part the response will have

(a) the system more heavily damped.
(b) the system damping gets reduced.
(c) time constant reduced.
(d) the system damping improved, time constant, rise time and the time at which the peak over-shoot occurs are all shortened.

29. For a prototype second order system if the pole is moved with a constant real part what will the response have in common?

(a) Time constant is increased.
(b) The damping ratio increase.
(c) Natural frequency of oscillation decrease.
(d) Damping ratio decrease, natural frequency of oscillation increases, time constant remains constant.

30. For a second order prototype system if the pole is moved along a radial line from the origin what effect the transient response will have?

(a) ω_n is decreased.
(b) Damping ratio is increased.
(c) Time constant, rise time and settling time are all increased.
(d) ω_n is increased, peak over short unchanged and the time constant, rise time and settling time are all decreased.

31. Consider a prototype second order system and a system included with a zero. The characteristic equation and ζ and ω_n are same for both the systems. What is the effect of time domain specification by addition of a zero?

(a) By applying the standard expressions obtained derived for the prototype system the specifications for both the systems are obtained which are same.
(b) Erroneous results will be obtained for some specifications if step (a) is followed.
(c) For the system with the addition of zero no formula is available and one has to proceed from first principles according to the required specifications.
(d) None of the above procedure is valid.

32. Consider the following forward path unity feedback system.

$$G(s) = \frac{3s + 13}{(s + 3)(s + 4)}$$

the ζ_e and ω_n are

(a) $\zeta = 0.1$ and $\omega_n = 5$ rad/sec.
(b) $\zeta = 1.0$ and $\omega_n = \sqrt{5}$ rad/sec.
(c) $\zeta = 1.0$ and $\omega_n = 5$ rad/sec.
(d) $\zeta = 0.5$ and $\omega_n = 4$ rad/sec.

33. Consider the following forward path T.F. of a certain unity feedback control systems.

$$G(s) = \frac{2(s+2)}{s(s+4)}$$

the ζ_e and ω_n are

(a) $\zeta = 3$ and $\omega_n = 4$ rad/sec.
(b) $\zeta = 1$ and $\omega_n = 2$ rad/sec.
(c) $\zeta = 2$ and $\omega_n = 2$ rad/sec.
(d) $\zeta = 1.5$ and $\omega_n = 2$ rad/sec.

34. Consider the following forward path T.F. of a certain unity feedback control system.

$$G(s) = \frac{(s+9)}{s(s+2)}$$

The nature of damping and ω_n are

(a) under damped and $\omega_n = 3$ rad/sec.
(b) critically damped and $\omega_n = 9$ rad/sec.
(c) over damped and $\omega_n = \sqrt{9/2}$ rad/sec.
(d) critically damped and $\omega_n = 3$ rad/sec.

35. A certain unity feedback control system has the following forward path T.F.

$$G(s) = \frac{(s+K)}{s(s+3)}$$

Find the value of K for $\varepsilon = 0.5$.

(a) $K = 8$.
(b) $K = 16$.
(c) $K = 36$.
(d) $K = 6$.

36. A certain control system has the following forward path T.F.

$$G(s) = \frac{e^{-3s}}{(s+2)(s+3)}$$

The impulse response of the above system is

(a) $g(t) = [e^{-2t} - e^{-3t}]u(t)$.
(b) $g(t) = [e^{-2t} + e^{-3t}]u(t-3)$.
(c) $g(t) = [e^{-2(t-3)} - e^{-3(t-3)}]u(t-3)$.
(d) $g(t) = [e^{-2(t+3)} - e^{-3(t+3)}]u(t+3)$.

37. Consider the following linear differential equation

$$\frac{d^2c}{dt^2} + 4\frac{dc}{dt} + 8c = 8r(t)$$

The damping ratio and damped frequency of oscillation are

(a) $\zeta = 0.707$ and $\omega_d = 2.83$ rad/sec.
(b) $\zeta = 0.707$ and $\omega_d = 2$ rad/sec.
(c) $\zeta = 1$ and $\omega_d = 0$
(d) $\zeta = 0.5$ and $\omega_d = 2.83$ rad/sec.

38. A certain unity negative feedback control system has the following forward path T.F. the input applied is step.

$$G(s) = \frac{K}{s^2 + s + 10}$$

The value of K for $e_{ss} < 0.5$ is,

(a) $K = 40$.
(b) $K = 30$.
(c) $K = 20$.
(d) $K = 10$.

39. A certain unity negative feedback has the following forward path T.F.

$$G(s) = \frac{(s+2)}{s^2(2s+3)(s+5)}$$

The input applied is

$$r(t) = \frac{3}{2}t^2$$

The steady state error is given by

(a) $e_{ss} = 10$.
(b) $e_{ss} = 15$.
(c) $e_{ss} = 20.5$.
(d) $e_{ss} = 22.5$.

40. Consider the closed loop T.F. of a certain control system

$$\frac{G(s)}{R(s)} = T(s) = \frac{(Ks + B)}{s^2 + As + B}$$

The steady error for unit ramp input is

(a) $e_{ss} = \frac{B}{(A+K)}$.
(b) $e_{ss} = \frac{B}{(A-K)}$.
(c) $e_{ss} = \frac{(A+K)}{B}$.
(d) $e_{ss} = \frac{(A-K)}{B}$.

41. For an unity negative feedback system $G(s)$ and $T(s) = C(s)/R(s)$ are connected by

(a) $G(s) = \frac{T(s)}{1-T(s)}$.
(b) $G(s) = \frac{T(s)}{1+T(s)}$.
(c) $G(s) = \frac{1-T(s)}{T(s)}$.
(d) $G(s) = \frac{1+T(s)}{T(s)}$.

42. The unit ramp response of a certain negative feedback system is given by

$$c(t) = (t - 1 + e^{-t} \cos t)u(t)$$

The forward path T.F. $G(s)$ is

(a) $G(s) = \frac{2}{s^2(s+2)}$.
(b) $G(s) = \frac{2}{s(s+2)}$.
(c) $G(s) = \frac{2}{(s+2)}$.
(d) $G(s) = \frac{2(s+1)}{(s+2)}$.

43. For a non-unity feedback system, if $G(s)$ is the forward T.F. and $H(s)$ is the T.F. of the feedback path, the equivalent unity feedback system has the following forward path T.F. $G_e(s)$.

(a) $G_e(s) = \frac{G(s)}{1+GH(s)}$.
(b) $G_e(s) = \frac{G(s)}{1-GH(s)}$.
(c) $G_e(s) = \frac{G(s)}{1+G(s)H(s)-G(s)}$.
(d) $G_e(s) = \frac{G(s)}{1+GH(s)+G(s)}$.

44. A certain feedback control system has the following T.F.

$$G(s) = \frac{5}{s(s+2)}$$

$$H(s) = \frac{1}{s+3}$$

The error is defined as $e(t) = r(t) - c(t)$. Find the e_{ss} for $r(t) = 2u(t)$.

(a) $e_{ss} = 2$.
(b) $e_{ss} = -2$.
(c) $e_{ss} = 4$.
(d) $e_{ss} = -4$.

45. For a certain feedback control system

$$G(s) = \frac{25}{s(s+5)}$$
$$H(s) = (1 + 0.076)$$

The input and error are defined as

$$r(t) = 4tu(t)$$
$$e(t) = r(t) - c(t)$$

The velocity error constants and the steady state error are

(a) $K_v = \frac{30}{5}$; $e_{ss} = \frac{2}{3}$ rad/sec.
(b) $K_v = \frac{25}{6}$; $e_{ss} = \frac{50}{3}$ rad/sec.
(c) $K_v = \frac{25}{6.9}$; $e_{ss} = 1.1$ rad/sec.
(d) $K_v = \frac{6.9}{25}$; $e_{ss} = \frac{100}{6.9}$ rad/sec.

46. The transfer function of a certain feedback control system is given by

$$T(s) = \frac{5}{(s^2 + 3s + 5)}$$
$$r(t) = 5tu(t)$$
$$e(t) = r(t) - c(t)$$

The steady state error is

(a) $e_{ss} = 2$ rad/sec.
(b) $e_{ss} = 5$ rad/sec.
(c) $e_{ss} = 8$ rad/sec.
(d) $e_{ss} = 3$ rad/sec.

47. For a certain non-unity feedback system

$$G(s) = \frac{K}{(s+3)}$$
$$H(s) = \frac{4}{(s+5)}$$

$$r(t) = tu(t)$$
$$e(t) = r(t) - c(t)$$

For what value of K the steady state error is zero.

(a) $K = 10$.
(b) $K = 15$.
(c) $K = 20$.
(d) $K = 25.5$.

48. For a certain non-unity feedback system

$$G(s) = \frac{K}{s(s+2)}$$
$$H(s) = (1 + 5s)$$
$$r(t) = tu(t)$$
$$e(t) = r(t) - c(t)$$

The steady state error is 6 rad/sec. The value of K is

(a) $e_{ss} = 8$.
(b) $e_{ss} = 10$.
(c) $e_{ss} = 2$.
(d) $e_{ss} = 1$.

49. A feedback control system is subjected to input $r(t)$ and disturbance $d(t)$ simultaneously. Under this condition,

(a) the characteristic equation for $r(t)$ and $d(t)$ disturbance will be same.
(b) characteristic equation does not depend upon input and disturbance.
(c) pole-zero cancellation takes place. New characteristic equation is formed.
(d) none of the above.

50. By using rate (tach-generator) feedback controller

(a) the damping is increased.
(b) the damping is decreased.
(c) steady state error is decreased.
(d) stability margin is decreased.

51. The use of proportional controller

(a) increases the overshoot.
(b) rise time settling time are increased.
(c) bandwidth is decreased.
(d) steady state error is decreased.

52. An integral (reset) controller

(a) improves the transient response.
(b) decrease the transient response.
(c) increases steady state error.
(d) order of the system is reduced.

53. A reset controller

(a) improves the transient response.
(b) reduces the type of the system.
(c) reduces the steady state error.
(d) increases the order of the system.

54. Derivative or rate controller

(a) improves the transient response.
(b) worsens the transient response.
(c) steady state error is decreased.
(d) order of the system is increased.

55. With the introduction of a rate controller

(a) rise time is reduced.
(b) settling time is reduced.
(c) peak over-shoot is reduced.
(d) all the above.

56. A PI controller

(a) increases the peak over-shoot.
(b) decreases the peak over-shoot.
(c) increases the bandwidth.
(d) decreases the relative stability.

57. A PI controller

(a) increases the peak over-shoot.
(b) phase margin and gain margin are reduced.
(c) reduces the steady state error.
(d) the type of the system is reduced by one.

58. A PD controller

(a) improves damping and reduces the peak-overshoot.
(b) increases the time response specifications.
(c) decreases the bandwidth.
(d) reduces the amplifier noise at higher frequencies.

59. Using a PID controller,

(a) stability is not guaranteed.
(b) good transient response is not guaranteed.
(c) steady state error is increased.
(d) by proper selection of PID controller parameters by using certain rules, stability and desired time domain specifications are achieved.

60. The ID controller is not used to control the process variables because

(a) the system steady state error is not satisfactory.
(b) the system transient response is not satisfactory.
(c) small variation of controller parameters makes the system unstable.
(d) all the above.

61. The transfer function of a certain first order system connected in a unity feedback system is

$$G(s) = \frac{4}{s+2}$$

An integral controller whose T.F

$$G(s) = \frac{4}{s}$$

is connected in cascade with the system. The % over-shoot for unit step input is

(a) $\% M_p = 24.43$.
(b) $\% M_p = 34.43$.
(c) $\% M_p = 44.43$.
(d) $\% M_p = 0$.

62. The forward path T.F. of a certain unity feedback system is

$$G(s) = \frac{1}{s(s+2)(s+3)}$$

A rate controller with the following T.F. is connected in cascade with it.

$$G(s) = T_d s$$

The value of T_d for the system to be critically damped is

(a) $T_d = 0.25$.
(b) $T_d = 0.5$.
(c) $T_d = 0.75$.
(d) $T_d = 1.0$.

63. A certain unity feedback system has the following T.F.

$$G(s) = \frac{36}{s(s+4)}$$

A PD controller is connected in cascade. The values of K_p and K_d for $\varepsilon = 0.8$ and $\omega_n = 9$ rad/sec are

(a) $K_p = 2; K_d = 0.9.$
(b) $K_p = 1.75; K_d = 0.53.$
(c) $K_p = 2.25; K_d = 0.29.$
(d) $K_p = 3.8; K_d = 0.07.$

64. A certain unity feedback control system has the following forward path T.F.

$$G(s) = \frac{10}{(s+a)}$$

An integral controller $G_c(s)$ is connected in cascade. The required e_{ss} for unit step input is 0.1 and the required $\omega_n = 12$ rad/sec. The values of a and K_I are

(a) $a = 10.5; K_I = 15.5.$
(b) $a = 12.0; K_I = 21.0.$
(c) $a = 14.4; K_I = 14.4.$
(d) $a = 18.4; K_I = 7.2.$

65. The unit impulse response of a first order system is given by

(a) $c(t) = Ke^{at}.$
(b) $c(t) = Ke^{-at}.$
(c) $c(t) = Ke^{at} \sin at.$
(d) $c(t) = Ke^{-at} \sin at.$

66. The unit step response of a first order system is given by

(a) $c(t) = K[1 + e^{at}].$
(b) $c(t) = K[1 - e^{-at}].$
(c) $c(t) = K[1 + e^{-at}].$
(d) $c(t) = K[1 - e^{-at}].$

67. Consider the following first order system

$$G(s) = \frac{K}{(s+a)}$$

The time constant of the system is

(a) $T = a.$
(b) $T = \frac{1}{a}.$

(c) $T = (a - 1)$.
(d) $T = (a + 1)$.

68. For the first order system in question 67, the rise time for unit step input is given by

(a) $t_r = 1.1\frac{1}{a}$.
(b) $t_r = 2.2\frac{1}{a}$.
(c) $t_r = 3.3\frac{1}{a}$.
(d) $t_r = 4.4\frac{1}{a}$.

69. For the first order system in question 67, the time delay for step input is

(a) $t_d = 0.127T$.
(b) $t_d = 0.352T$.
(c) $t_d = 0.491T$.
(d) $t_d = 0.693T$.

70. For the first order system in question 67, the settling time for step input for 2% error tolerance is

(a) $t_s = 3\frac{1}{a}$.
(b) $t_s = 3.91\frac{1}{a}$.
(c) $t_s = 2.66\frac{1}{a}$.
(d) $t_s = 4.231\frac{1}{a}$.

71. For the first order system in question 67, for unit step input, the % overshoot is

(a) $\% M_p e^{-a\pi} \times 100$.
(b) $\% M_p e^{a\pi} \times 100$.
(c) $\% M_p e^{a} \times 100$.
(d) $\% M_p = 0$.

T is the time constant.

72. The expression for unit impulse response of a standard under damped second order system is given by

(a) $c(t) = \omega_n^2 t e^{-\omega_n t}$.
(b) $c(t) = \dfrac{\omega_n}{\sqrt{1-\zeta^2}} e^{-\zeta\omega_n t} \sin(\omega_n\sqrt{1-\zeta^2})t$.
(c) $c(t) = \dfrac{\omega_n}{2\sqrt{\zeta^2-1}} e^{-\zeta\omega_n t}[\sin(\omega_n\sqrt{1-\zeta^2})t]$.
(d) $c(t) = \dfrac{\omega_n}{\sqrt{1-\zeta^2}} e^{\zeta\omega_n t} \sin(\omega_n\sqrt{1-\zeta^2})t$.

73. The expression for unit impulse response of a critically damped standard second order system is given by

(a) $c(t) = \omega_n^2 t e^{-\omega_n t}$.
(b) $c(t) = \omega_n^2 t e^{\omega_n t}$.
(c) $c(t) = \dfrac{\omega_n}{\sqrt{1-\zeta^2}} e^{-\zeta\omega_n t} \sin(\omega_n \sqrt{1-\zeta^2} t$.
(d) $c(t) = \dfrac{\omega_n}{\sqrt{1-\zeta^2}} e^{-\zeta\omega_n t} \sin \omega_n \sqrt{1-\zeta^2} t$.

74. The expression for unit step response of a standard critically damped second order system is given by

(a) $c(t) = \dfrac{1}{\omega_n^2}[1 - t\omega_n e^{-\omega_n t} - e^{-\omega_n t}]$.
(b) $c(t) = 1 - e^{-\omega_n t}[\sin \omega_n^2 t + \phi]$.
(c) $c(t) = \dfrac{1}{\omega_n^2}[1 + t\omega_n e^{-\omega_n t} - e^{-\omega_n t}]$.
(d) $c(t) = \omega_n^2 t e^{-\omega_n t}$.

75. The unit step response of a standard under damped second order system is given by

(a) $c(t) = \left[1 - \frac{e^{-\omega_n t}}{\sqrt{1-\zeta^2}} \left(\sin \omega_n \sqrt{1-\zeta^2} t + \phi\right)\right]$.
(b) $c(t) = \left[1 + \frac{e^{-\omega_n t}}{\sqrt{1-\zeta^2}} \left(\sin \omega_n \sqrt{1-\zeta^2} t + \phi\right)\right]$.
(c) $c(t) = \left[1 - \frac{e^{\omega_n t}}{\sqrt{1-\zeta^2}} \left(\sin \omega_n \sqrt{1-\zeta^2} t + \phi\right)\right]$.
(d) $c(t) = \left[1 + \frac{e^{-\omega_n t}}{\sqrt{1-\zeta^2}} \left(\sin \omega_n \sqrt{1-\zeta^2} t + \phi\right)\right]$.

76. The percentage peak over-shoot of a under damped standard second order system subjected to step input is

(a) $\% M_p = e^{\frac{\zeta\pi}{\sqrt{1-\zeta^2}}} \times 100$.
(b) $\% M_p = e^{\frac{-\zeta\pi}{\sqrt{1-\zeta^2}}} \times 100$.
(c) $\% M_p = e^{\frac{-\omega_n\pi}{\sqrt{1-\zeta^2}}} \times 100$.
(d) $\% M_p = e^{\frac{\omega_n\pi}{\sqrt{1-\zeta^2}}} \times 100$.

77. The time at which the maximum peak over-shoot occurs for a under damped standard second order system when subjected to step input is

(a) $t_\phi = \dfrac{2\pi}{\omega_d}$.
(b) $t_\phi = \dfrac{1}{\omega_d}$.

(c) $t_\phi = 2\pi\omega_d$.
(d) $t_\phi = \dfrac{\pi}{\omega_d}$.

78. A standard second order system is subjected to step input. The system is under damped. The under-shoots in the response occur at

(a) $t_u = \frac{n\pi}{\omega_d}$ where n is odd.
(b) $t_u = \frac{(n+1)\pi}{\omega_d}$ where n is odd.
(c) $t_u = \frac{n\pi}{\omega_d}$ where n is even.
(d) $t_u = \frac{(n+1)}{\omega_d}$ where n is even.

79. The time constant of a standard second order system is

(a) $T = \zeta\omega_n$.
(b) $T = \dfrac{1}{\zeta\omega_n}$.
(c) $T = \dfrac{\zeta}{\omega_n}$.
(d) $T = \dfrac{\omega_n}{\zeta}$.

80. A standard second order is given a step input. The settling time for 2%, 5% and 7% steady state error tolerance is respectively

(a) $t_s = 3.91T;\ 3T;\ 2.66T$.
(b) $t_s = 2.66T;\ 3T;\ 3.91T$.
(c) $t_s = 3T;\ 2.66T;\ 3.19T$.
(d) $t_s = 3T;\ 3.19T;\ 2.66T$.

where $T = \frac{1}{\zeta\omega_n}$ = time constant.

81. The standard second order system is given a step input. The rise time is given by

(a) $t_s = \dfrac{(\pi+\theta)}{\omega_d}$.
(b) $t_s = \dfrac{(\pi-\theta)}{\omega_d}$.
(c) $t_s = \dfrac{\omega_d}{(\pi+\theta)}$.
(d) $t_s = \dfrac{\omega_d}{(\pi-\theta)}$.

where $\omega_d = \omega_n\sqrt{1-\zeta^2}$ and $\tan\theta = \frac{\sqrt{1-\zeta^2}}{\zeta}$.

82. An under damped standard second order system is given a step input. The period of oscillation of the transient response is given by

(a) $T_0 = \dfrac{\omega_d}{2\pi}$.
(b) $T_0 = 2\pi\omega_d$.

(c) $T_0 = \frac{2\pi}{\omega_d}$.

(d) $T_0 = \frac{1}{2\pi\omega_d}$.

83. A certain closed loop system has the following forward path T.F.

$$G(s) = \frac{10}{s(s+2)}$$

The system is subjected to unit step input. The % over-shoot is given by

(a) $M_p = 25\%$.
(b) $M_p = 30\%$.
(c) $M_p = 32.5\%$.
(d) $M_p = 35\%$.

84. For the system in question 83, the rise time is given by

(a) $t_r = 0.43$ sec.
(b) $t_r = 0.51$ sec.
(c) $t_r = 0.63$ sec.
(d) $t_r = 0.72$ sec.

85. For the system described in question 83, the time constant is

(a) $T = 0.25$ sec.
(b) $T = 1.0$ sec.
(c) $T = 1.5$ sec.
(d) $T = 2.0$ sec.

86. For the system described in question 83, the settling time for 2% error tolerance is

(a) $t_s = 2.15$ sec.
(b) $t_s = 3.91$ sec.
(c) $t_s = 4.58$ sec.
(d) $t_s = 5.66$ sec.

87. For the system described in question 83, the period of oscillation is given by

(a) $T_0 = 2.09$ sec.
(b) $T_0 = 3.91$ sec.
(c) $T_0 = 5.33$ sec.
(d) $T_0 = 7.23$ sec.

88. A certain unity feedback control system has the following forward path T.F.

$$G(s) = \frac{(3s+13)}{(s+3)(s+4)}$$

The damping and natural frequency of oscillations are

(a) $\zeta = 0.8;\ \omega_n = 4$ rad/sec.
(b) $\zeta = 0.9;\ \omega_n = 3.2$ rad/sec.
(c) $\zeta = 1.0;\ \omega_n = 5$ rad/sec.
(d) $\zeta = 0.5;\ \omega_n = 2$ rad/sec.

89. A certain unity negative feedback control system has the following forward path T.F.

$$G(s) = \frac{(s + K)}{s(s + 3)}$$

Find the value of K so that $\zeta = 0.5$

(a) $K = 4$.
(b) $K = 8$.
(c) $K = 12$.
(d) $K = 16$.

90. Consider the following forward path T.F. of a certain unity negative feedback system.

$$G(s) = \frac{(s + K)}{s(s + 3)}$$

The required $\omega_n = 4$ rad/sec. The values of K and ζ are

(a) $K = 11.75;\ \zeta = 1$.
(b) $K = 10;\ \zeta = 1.7188$.
(c) $K = 11.75;\ \zeta = 2$.
(d) $K = 11.75;\ \zeta = 1.7188$.

91. The open loop T.F. of a certain unity negative feedback system is

$$G(s) = \frac{2(s + 8)}{s(s + 2)}$$

The unit step response is given by

(a) $c(t) = 1 - e^{-t}\sin 5t$.
(b) $c(t) = 1 - e^{-2t}\sin 5t$.
(c) $c(t) = 1 - e^{-2t}\cos 3.464t$.
(d) $c(t) = 1 - e^{-2t}\sin 3.464t$.

92. Consider the system shown in Question 91. The % peak over-shoot is given by

(a) $\%\ M_p = 19.1$.
(b) $\%\ M_p = 17.51$.
(c) $\%\ M_p = 21.26$.
(d) $\%\ M_p = 2.23$.

93. For the system given in Question 91, the rise time is given by

(a) $t_r = 0.3525$ sec.
(b) $t_r = 0.1256$ sec.
(c) $t_r = 1.025$ sec.
(d) $t_r = 0.4533$ sec.

94. Consider the following unit step response of a negative unity feedback system.

$$c(t) = 10 - 30e^{-2t} + 20e^{-3t}$$

Determine the forward path T.F.

(a) $G(s) = \frac{6}{s(s+6)}$.
(b) $G(s) = \frac{9}{s(s+9)}$.
(c) $G(s) = \frac{4}{s(s+4)}$.
(d) $G(s) = \frac{6}{s(s+5)}$.

95. For the system described in Question 94, find the time constant.

(a) $T = 0.2$ sec.
(b) $T = 0.4$ sec.
(c) $T = 0.6$ sec.
(d) $T = 1.12$ sec.

96. A certain unity negative feedback control system has the following forward path T.F.

$$G(s) = \frac{5G_e(s)}{s^2}$$

Find the T.F. $Ge(s)$ **to give** $\zeta = 0.5$ **and** $\omega_n = 5$ **rad/sec.**

(a) $G_e(s) = \frac{1}{(s+2)}$.
(b) $G_e(s) = (s+2)$.
(c) $G_e(s) = \frac{1}{(s+4)}$.
(d) $G_e(s) = (s+5)$.

97. The % over-shoot of a certain unity negative feedback system is $M_p = 50\%$ **when subjected to step unit. The damping ratio of the system is**

(a) $\zeta = 0.125$.
(b) $\zeta = 0.215$.
(c) $\zeta = 0.515$.
(d) $\zeta = 0.707$.

98. For a certain system with T.F. $G(s)$ **the step response is given by**

$$g(t) = \left[1 - \frac{3}{2}e^{-2t} + e^{-3t}\right]u(t)$$

What will be the time response for $G(s)e^{-2s}$

(a)

$$g(t) = \left[1 - \frac{3}{2}e^{-2t-2} + e^{-3t-2}\right] u(t-2).$$

(b)

$$g(t) = \left[1 - \frac{3}{2}e^{-2t+2} + e^{-3t+2}\right] u(t+2).$$

(c)

$$g(t) = \left[1 - \frac{3}{2}e^{-2(t-2)} + e^{-3(t-2)}\right] u(t-2).$$

(d)

$$g(t) = \left[1 - \frac{3}{2}e^{-2(t+2)} + e^{-3(t+2)}\right] u(t+2).$$

99. A certain unity negative feedback control system has the following open loop T.F.

$$G(s) = \frac{K}{(s+A)(s+2)}$$

Find the value of K and A for $\zeta = 0.707$ and $t_p = 1.8$ sec.

(a) $A = 1.1;\ K = 3.1.$
(b) $A = 1.49;\ K = 3.125.$
(c) $A = 1.236;\ K = 3.822.$
(d) $A = 1.5;\ K = 3.0.$

100. The forward path T.F of a certain unity feedback control system is given by

$$G(s) = \frac{(s+a)}{(s+b)^2}$$

The desired closed loop T.F. is

$$\frac{G(s)}{R(s)} = \frac{(s+96)}{(s^2+11s+121)}$$

The values of a and b are

(a) $a = 9.6;\ b = 11.$
(b) $a = 96;\ b = 6.$
(c) $a = 6;\ b = 96.$
(d) $a = 96;\ b = 5.$

101. A certain unity feedback control system has the following forward path T.F. $G(s)$. The system is subjected to step input of strength R. What is position error constant and the steady state error?

(a) $K_p = \lim_{s\to 0} G(s);\ e_{ss} = \frac{R}{K_p}$.
(b) $K_p = \lim_{s\to 0} sG(s);\ e_{ss} = \frac{R}{1+K_p}$.
(c) $K_p = \lim_{s\to 0} G(s);\ e_{ss} = \frac{R}{1+K_p}$.
(d) $K_p = \lim_{s\to 0} G(s);\ e_{ss} = \frac{1}{1+K_p}$.

102. A certain unity negative feedback system has $G(s)$ as the forward path T.F. The impact has a slope of R rad/sec. The steady error and velocity error constants are given by

(a) $K_v = \lim_{s\to 0} G(s);\ e_{ss} = \frac{R}{K_v}$.
(b) $K_v = \lim_{s\to 0} sG(s);\ e_{ss} = \frac{R}{1+K_v}$.
(c) $K_v = \lim_{s\to 0} s^2G(s);\ e_{ss} = \frac{R}{1-K_v}$.
(d) $K_v = \lim_{s\to 0} sG(s);\ e_{ss} = \frac{R}{K_v}$.

103. A certain unity negative feedback system that has the forward path T.F. is $G(s)$. The input $r(t) = \frac{R}{2}t^2$. The acceleration error coefficient and the steady state error are given by

(a) $K_a = \lim_{s\to 0} s^2G(s);\ e_{ss} = \dfrac{R}{K_a}$.
(b) $K_a = \lim_{s\to 0} s^2G(s);\ e_{ss} = \dfrac{R}{1+K_a}$.
(c) $K_a = \lim_{s\to 0} sG(s);\ e_{ss} = \dfrac{R}{K_a}$.
(d) $K_a = \lim_{s\to 0} s^2G(s);\ e_{ss} = \dfrac{R}{1+K_a}$.

104. A certain unity negative feedback system has the following closed loop T.F.

$$\frac{c(s)}{R(s)} = \frac{(Ks + B)}{(s^2 + As + B)}$$

The steady state error for unit ramp input is

(a) $e_{ss} = \dfrac{A+K}{B}$.
(b) $e_{ss} = \dfrac{A}{B+K}$.
(c) $e_{ss} = \dfrac{A+B}{K}$.
(d) $e_{ss} = \dfrac{A-K}{B}$.

105. A certain unity feedback control system when subjected to unit ramp input has the following response.

$$c(t) = t - 1 + e^{-t}\cos t$$

Find the steady state error and the open loop T.F.

(a) $e_{ss} = 2;\ G(s) = \dfrac{2}{s(s+2)}$.
(b) $e_{ss} = 1;\ G(s) = \dfrac{2}{s(s+1)}$.
(c) $e_{ss} = 1;\ G(s) = \dfrac{2}{s(s+2)}$.
(d) $e_{ss} = 2;\ G(s) = \dfrac{1}{s(s+1)}$.

106. A certain non-unity feedback control system has the following forward path and feedback path T.F.s.

$$G(s) = \frac{5}{s(s+2)}$$
$$H(s) = \frac{1}{(s+3)}$$

Find the steady state error for unit step input. $e(t) = r(t) - c(t)$.

(a) $e_{ss} = -4$.
(b) $e_{ss} = +4$.
(c) $e_{ss} = -2$.
(d) $e_{ss} = +2$.

107. A certain unity feedback control system has the following T.F.s. connected in cascade.

$$G_1(s) = \frac{10}{(s+2)}$$
$$G_2(s) = \frac{5}{(s+3)(s+6)}$$

The system is subjected to a step input of strength 3 and a step disturbance of strength 2 applied between $G_1(s)$ **and** $G_2(s)$**. Find the steady state error.**

(a) $e_{ss} = 0.53$.
(b) $e_{ss} = 1.03$.
(c) $e_{ss} = 3.24$.
(d) $e_{ss} = 4.59$.

108. A certain unity feedback system has the following forward path T.F.

$$G(s) = \frac{5}{(s+5)}$$

The input applied is $r(t) = 6 + 5t$. The generalized dynamic error coefficients are

(a) $c_0 = 0.5;\ c_1 = 0.05.$
(b) $c_0 = 0.05;\ c_1 = 0.5.$
(c) $c_0 = 50;\ c_1 = 5.$
(d) $c_0 = 5;\ c_1 = 50.$

109. A certain non-unity negative feedback control system has the forward path T.F.

$$G(s) = \frac{2}{(s+1)(s+2)}$$

and the feedback T.F. $H(s) = (s+3)$. The equivalent unity feedback system will have a forward path T.F. as

(a)

$$G_e(s) = \frac{4}{(s^2+5s+6)}.$$

(b)

$$G_e(s) = \frac{2}{(s^2+5s+6)}.$$

(c)

$$G_e(s) = \frac{8}{(s^2+5s+6)}.$$

(d)

$$G_e(s) = \frac{15}{(s^2+5s+6)}.$$

110. A certain control system has the following closed loop T.F.

$$\frac{C(s)}{R(s)} = T(s) = \frac{5}{(s^2+3s+5)}$$

The steady state error for ramp input of slope 5 red/sec is

(a) $e_{ss} = \frac{5}{3}.$
(b) $e_{ss} = 5.$

(c) $e_{ss} = 3$.
(d) $e_{ss} = 15$.

111. The forward path and feedback path T.F. of a negative feedback system are

$$G(s) = \frac{K}{(s+3)}$$
$$H(s) = \frac{4}{(s+5)}$$

Find the value of K so that the steady state error is zero for ramp input.

$$e(t) = r(t) - c(t)$$

(a) $K = 5$.
(b) $K = 10$.
(c) $K = 15$.
(d) $K = 20$.

112. A certain non-unity feedback control system has the following T.F.s

$$G(s) = \frac{5}{s(s+2)}$$
$$H(s) = \frac{1}{(s+5)}$$

The system is subjected to unit step input. The steady state actuating signal is

(a) $e_a(t) = -2$.
(b) $e_a(t) = -4$.
(c) $e_a(t) = \infty$.
(d) $e_a(t) = 0$.

113. Consider the following closed loop T.F. of a certain negative feedback system.

$$T(s) = \frac{(s-6)}{s^2 + 10s + 16}$$

Find the damping ratio and nature of damping.

(a) $\zeta = 0.707$ under damped.
(b) $\zeta = 1.0$ critically damped.
(c) $\zeta = 1.25$ over damped.
(d) $\zeta = 0$ undamped.

114. A certain control system has the forward path T.F. as

$$G(s) = \frac{(25 - Ts)}{s(s + 10)}$$

Find the value of T so that the system is undamped.

(a) $T = 2$.
(b) $T = 5$.
(c) $T = 8$.
(d) $T = 10$.

115. For the following closed loop T.F. of a certain system find the value of K so that the damping ratio $\zeta = 1$.

$$T(s) = \frac{(s + 2)}{5s^2 + 2s + K}$$

(a) $K = 0.1$.
(b) $K = 0.2$.
(c) $K = 0.4$.
(d) $K = 8$.

116. For the following closed loop T.F. find the value of K so that $\omega_d = 2$ rad/sec.

$$T(s) = \frac{(Ks + 2)}{5s^2 + (2K + 5)s + 25}$$

(a) $K = 0.5$.
(b) $K = 1.0$.
(c) $K = 2.0$.
(d) $K = 2.5$.

117. For the following closed loop T.F. find the value of K so that $\omega_n = 8$ rad/sec.

$$T(s) = \frac{(s + K)}{Ks^2 + 5s + 100}$$

(a) $K = 0.985$.
(b) $K = 1.563$.
(c) $K = 2.122$.
(d) $K = 2.736$.

118. For the closed loop T.F. given below find the % over-shoot for unit step input

$$T(s) = \frac{125}{(5s^2 + 10s + 500)}$$

(a) $\% M_p = 66.66$.
(b) $\% M_p = 72.94$.
(c) $\% M_p = 76.55$.
(d) $\% M_p = 79.89$.

119. A certain closed loop system has the following T.F.

$$T(s) = \frac{64}{(s^2 + 8s + 64)}$$

The system is subjected to unit step input. Find t_p

(a) $t_p = 0.4535$ sec.
(b) $t_p = 0.5231$ sec.
(c) $t_p = 1.2531$ sec.
(d) $t_p = 1.5112$ sec.

120. A certain system has the following closed loop T.F.

$$T(s) = \frac{100}{5s^2 + 10s + 100}$$

The system is subjected to unit step input. Find the rise time t_r

(a) $t_r = 0.333$ sec.
(b) $t_r = 0.459$ sec.
(c) $t_r = 0.412$ sec.
(d) $t_r = 0.295$ sec.

Answers

(1) c
(2) a
(3) c
(4) a
(5) d
(6) d
(7) a
(8) b
(9) b
(10) d
(11) c
(12) b
(13) a
(14) c
(15) a
(16) c
(17) b
(18) c
(19) c
(20) a
(21) b
(22) c
(23) d
(24) d
(25) a
(26) a
(27) b
(28) d
(29) d
(30) d
(31) c
(32) c
(33) d
(34) a
(35) b
(36) c
(37) b
(38) b
(39) d
(40) b
(41) a
(42) b
(43) c
(44) d
(45) c
(46) d
(47) b
(48) c
(49) b
(50) a
(51) a
(52) b
(53) c
(54) a
(55) d
(56) b
(57) c
(58) a
(59) d
(60) c
(61) c
(62) a
(63) c
(64) c
(65) b

(66) d	(77) d	(88) c	(99) b	(110) c
(67) b	(78) c	(89) d	(100) d	(111) c
(68) b	(79) b	(90) d	(101) c	(112) d
(69) d	(80) a	(91) c	(102) d	(113) c
(70) b	(81) b	(92) a	(103) a	(114) d
(71) d	(82) c	(93) d	(104) d	(115) b
(72) b	(83) d	(94) d	(105) c	(116) d
(73) a	(84) c	(95) b	(106) c	(117) b
(74) a	(85) b	(96) d	(107) b	(118) b
(75) a	(86) b	(97) b	(108) a	(119) a
(76) b	(87) a	(98) c	(109) b	(120) c

Chapter 3
Frequency Response Analysis

Multiple Choice Questions

Choose the Correct Answer

1. **The frequency response is a**
   (a) transient response.
   (b) steady state response.
   (c) transient and steady state response.
   (d) an impulse response.

2. **The components of frequency response are**
   (a) magnitude component.
   (b) phase component.
   (c) magnitude and phase component.
   (d) frequency component.

3. **The frequency response of a system with T.F. $G(s)$ is**
   (a) variation of $|G(j\omega)|$ versus ω.
   (b) variation of $\angle G(j\omega)$ versus ω.
   (c) variation of $|G(j\omega)|$ and $\angle G(j\omega)$ versus ω.
   (d) All the above

© The Author(s), under exclusive license to Springer Nature Switzerland AG 2026
S. Palani, *Multiple-Choice Questions in Control Systems Engineering*,
https://doi.org/10.1007/978-3-031-93787-3_3

4. The sinusoidal T.F. is

(a) obtained as $G(j\omega)$ by substituting $s = j\omega$ in $G(s)$.
(b) the Laplace inverse of $g(t)$.
(c) the Laplace inverse of $G(s)$ when step input is applied.
(d) all the above.

5. The advantage of frequency response method is

(a) it can be applied to both linear as well as non-linear systems.
(b) the frequency response tests are done under transient condition which are simpler for analysis.
(c) it is easy to generate sine wave and collect data to draw frequency response which can be used for analysis and design.
(d) in frequency response method the analysis is done at one frequency at a time which requires less test time.

6. The unique disadvantage of frequency response method is

(a) it requires sinusoidal wave under transient condition which is difficult to generate.
(b) it is not easy to correlate the frequency domain specifications to time domain specifications (second order system is an exemption).
(c) it can be applied to all non-linear systems.
(d) measurement of data under transient condition is difficult.

7. The T.F. of a certain system is given by

$$G(s) = \frac{1}{(s+4}$$

The input $r(t) = 5\sin(2t + 20°)$ **the output is**

(a) $y(t) = 5\sin(2t + 63.4°)$
(b) $y(t) = 5\sin(2t - 63.4°)$
(c) $y(t) = 1.18\sin(2t + 43.4°)$
(d) $y(t) = 1.18\sin(2t - 43.4°)$

8. The magnitude and phase angle of a prototype second order system at $\omega = \omega_n$ **is**

(a)

$$G(j\omega) = \frac{1}{\zeta}\,\underline{\lfloor -90°}$$

(b)

$$G(j\omega) = \frac{1}{\zeta}\,\underline{\lfloor 90°}$$

(c)

$$G(j\omega) = \frac{1}{2\zeta} \underline{|-90^\circ}$$

(d)

$$G(j\omega) = \frac{1}{2\zeta} \underline{|180^\circ}$$

9. The unique advantage of polar plot is

(a) the magnitude response is displayed single plot.
(b) the phase angle plot is displayed single plot.
(c) the magnitude as well as phase angle plot are displayed in a single plot over a wide frequency range.
(d) the famous Nyquist stability criterion uses polar plot.

10. For the following transfer function $G(s)$, the sinusoidal T.F. $G(j\omega)$ has the phase angle,

$$G(s) = \frac{K(s-10)}{s(-s+5)}$$

(a)

$$\underline{|G(j\omega)} = \underline{|\tan^{-1}\frac{\omega}{5} - \tan^{-1}\frac{\omega}{10}}$$

(b)

$$\underline{|G(j\omega)} = \underline{|90^\circ - \tan^{-1}\frac{\omega}{10} - \tan^{-1}\frac{\omega}{5}}$$

(c)

$$\underline{|G(j\omega)} = \underline{|-90^\circ - \tan^{-1}\frac{\omega}{10} + \tan^{-1}\frac{\omega}{5}}$$

(d)

$$\underline{|G(j\omega)} = \underline{|-90^\circ + \tan^{-1}\frac{\omega}{10} - \tan^{-1}\frac{\omega}{5}}$$

11. For the following T.F., the sinusoidal T.F. has the phase angle,

$$G(s) = \frac{k(s+10)}{(s^2+4s+16)}$$

(a)

$$\underline{|G(j\omega)} = \tan^{-1}\frac{\omega}{10} + \tan^{-1}\frac{4\omega}{(16-\omega^2)}.$$

(b)

$$\angle G(j\omega) = \tan^{-1}\frac{\omega}{10} - \tan^{-1}\frac{4\omega}{(16-\omega^2)}.$$

(c)

$$\angle G(j\omega) = \tan^{-1}\frac{\omega}{10} - \tan^{-1}\frac{4\omega}{(16+\omega^2)}.$$

(d)

$$\angle G(j\omega) = \tan^{-1}\frac{10}{\omega} + \tan^{-1}\frac{\omega}{(16+\omega^2)}.$$

12. For the following T.F., the sinusoidal T.F. has the phase angle component of

$$G(s) = \frac{k}{(s-10)}$$

(a)

$$\angle G(j\omega) = \tan^{-1}\frac{\omega}{10}.$$

(b)

$$\angle G(j\omega) = -\tan^{-1}\frac{\omega}{10}.$$

(c)

$$\angle G(j\omega) = 180 + \tan^{-1}\frac{\omega}{10}.$$

(d)

$$\angle G(j\omega) = 180 - \tan^{-1}\frac{\omega}{10}.$$

13. Consider the following T.F.

$$G(s) = \frac{k}{(10-s)}$$

The phase angle component of $G(j\omega)$ is

(a)

$$\angle G(j\omega) = \tan^{-1}\frac{\omega}{10}.$$

(b)

$$\angle G(j\omega) = -\tan^{-1}\frac{\omega}{10}.$$

(c)

$$\underline{|G(j\omega)} = 180 + \tan^{-1}\frac{\omega}{10}.$$

(d)

$$\underline{|G(j\omega)} = 180 - \tan^{-1}\frac{\omega}{10}.$$

14. Consider the following T.F.

$$G(s) = \frac{-k(s+5)}{s^2(s-5)}$$

The phase component of $G(j\omega)$ **is,**

(a)

$$\underline{|G(j\omega)} = \tan^{-1}\frac{\omega}{5}.$$

(b)

$$\underline{|G(j\omega)} = -\tan^{-1}\frac{\omega}{5}.$$

(c)

$$\underline{|G(j\omega)} = 180^\circ - \tan^{-1}\frac{\omega}{5}.$$

(d)

$$\underline{|G(j\omega)} = 180^\circ + 2\tan^{-1}\frac{\omega}{5}.$$

15. For the $G(j\omega) = K(s+10)$, **the polar plot is** (Fig. 3.1).
16. For the $G(j\omega) = K(s-5)$, **the polar plot is** (Fig. 3.2).
17. For the $G(s) = (-s+5)$, **the polar plot is** (Fig. 3.3).
18. For the $G(s) = K/(s-10)$, **sketch polar plot** (Fig. 3.4).
19. For the $G(s) = K/(-s+10)$, **sketch polar plot** (Fig. 3.5).
20. Draw the polar plot for $G(s) = K(10-s)/(s-5)$ (Fig. 3.6).
21. For $G(s) = K(10-s)/s(s-5)$, **draw the polar plot** (Fig. 3.7).
22. Draw the polar plot for $G(s) = K(s-2)/(s+2)$ (Fig. 3.8).
23. The frequency response of a type zero minimum phase system starts ($\omega = 0$) **from**

(a) positive real axis.
(b) negative real axis.
(c) ∞ in the positive imaginary axis.
(d) ∞ in the negative imaginary axis.

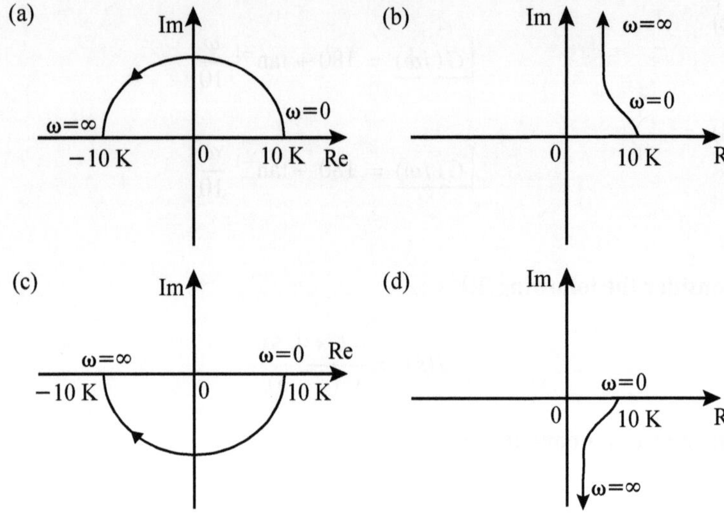

Fig. 3.1 Polar plot for $G(j\omega) = K(s + 10)$

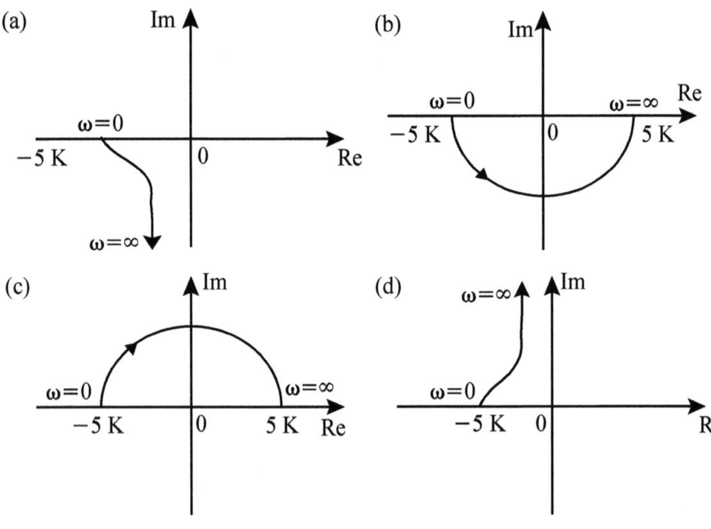

Fig. 3.2 Polar plot for $G(j\omega) = K(s - 5)$

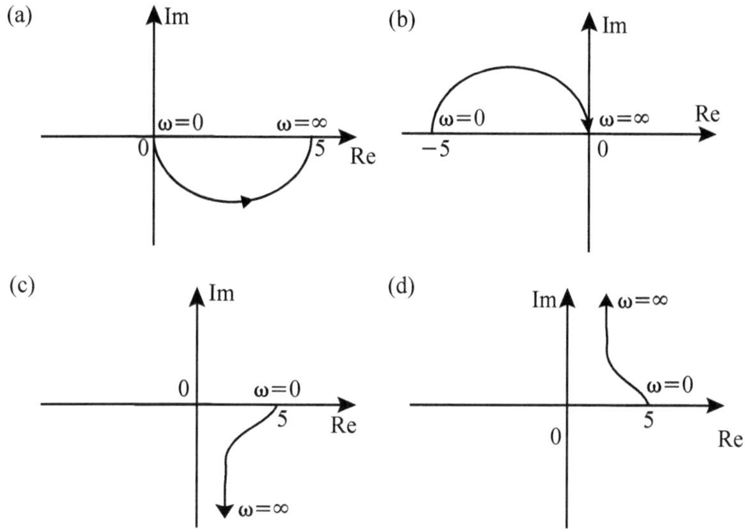

Fig. 3.3 Polar plot for $G(s) = (-s + 5)$

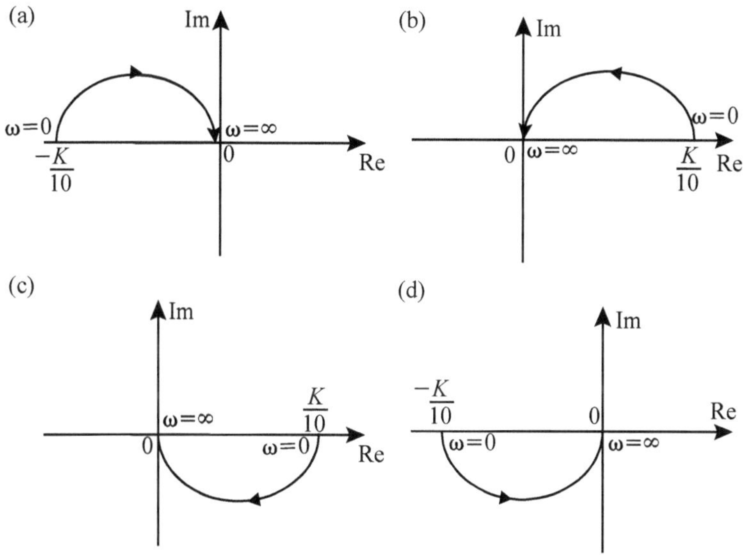

Fig. 3.4 Polar plot for $G(s) = K/(s - 10)$

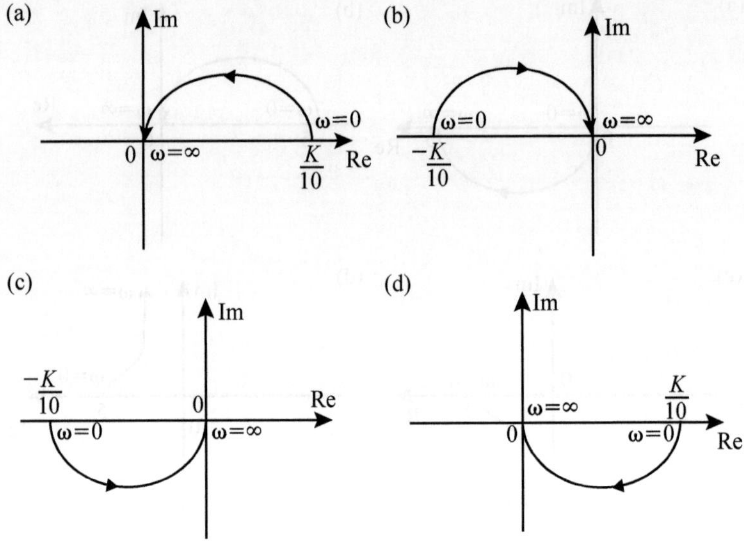

Fig. 3.5 Polar plot for $G(s) = K/(-s + 10)$

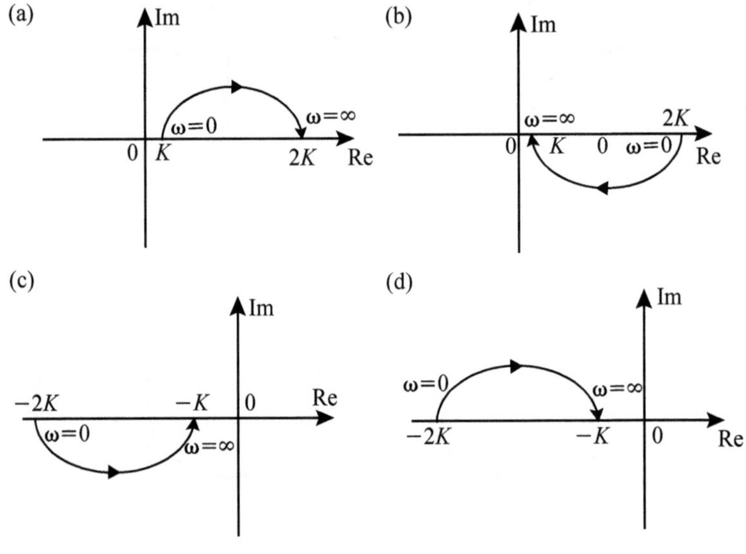

Fig. 3.6 Polar plot for $G(s) = K(10 - s)/(s - 5)$

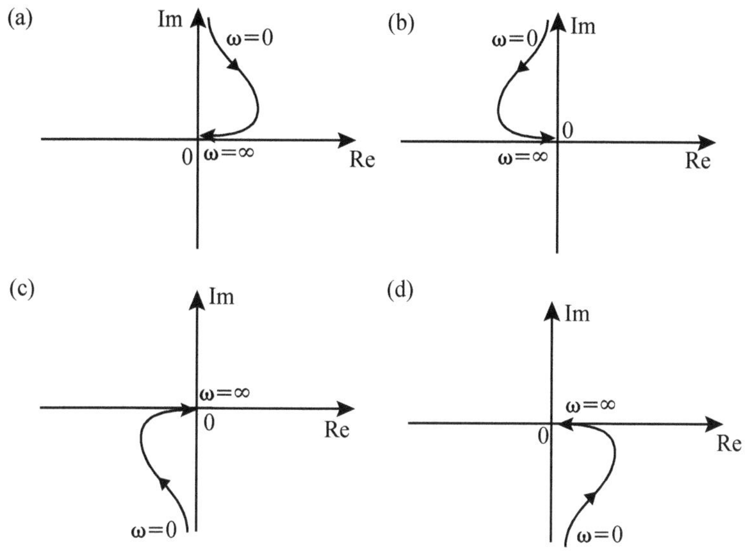

Fig. 3.7 Polar plot for $G(s) = K(10 - s)/s(s - 5)$

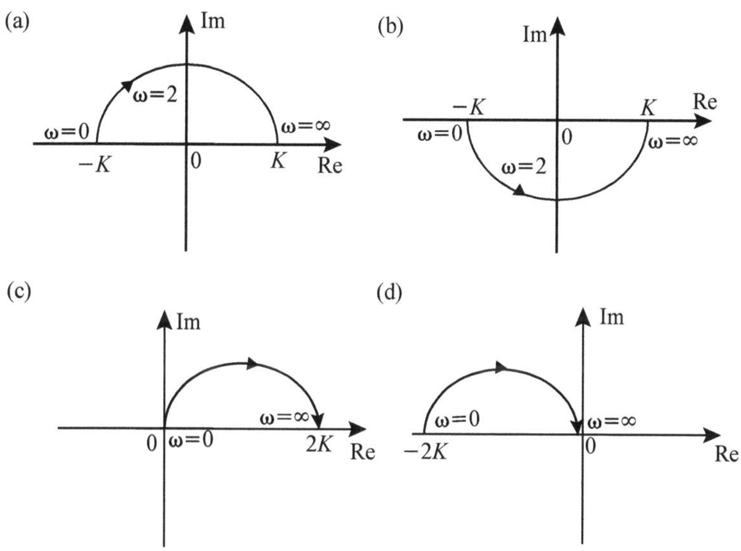

Fig. 3.8 Polar plot for $G(s) = K(s - 2)/(s + 2)$

24. The frequency response of a type one minimum phase system starts ($\omega = 0$) from

(a) positive real axis.
(b) negative real axis.
(c) ∞ in the positive imaginary axis.
(d) ∞ in the negative imaginary axis.

25. The frequency response of a type two minimum phase system starts ($\omega = 0$) from

(a) positive real axis.
(b) ∞ from negative real axis.
(c) ∞ in the positive real axis.
(d) ∞ in the negative real axis.

26. The frequency response of a minimum phase first order system (without any zero) ends ($\omega = \infty$) at the

(a) origin in the fourth quadrant.
(b) origin in the third quadrant.
(c) origin in the second quadrant.
(d) origin in the first quadrant.

27. The frequency response of a minimum phase second order system (without any zero) ends ($\omega = \infty$) at the

(a) origin in the first quadrant.
(b) origin in the second quadrant.
(c) origin in the third quadrant.
(d) origin in the fourth quadrant.

28. The frequency response of a third order system of minimum phase (with any zero) ends ($\omega = \infty$) at the

(a) origin in the fourth quadrant.
(b) origin in the third quadrant.
(c) origin in the second quadrant.
(d) origin in the first quadrant.

29. The Bode frequency response plot is

(a) the plot of $\angle G(j\omega)$ vs. $\log \omega$.
(b) the plot of $\angle G(j\omega)$ vs. $\log \omega$.
(c) the plot of $\angle G(j\omega)$ and $20 \log |G(j\omega)|$ vs. $\log \omega$.
(d) the plot of $|G(j\omega)|$ vs. $\angle G(j\omega)$

30. The unique advantage of Bode plot is

(a) The vertical axis represents the magnitude in dB.
(b) The horizontal axis represents the frequency in log scale.
(c) Information about the magnitude and phase are known for a wide frequency range.
(d) Multiplication and division becomes addition and subtraction in Bode plot.

31. The unique disadvantage of Bode plot is

(a) Bode magnitude plot is sketched using straight line approximation which introduces error.
(b) since the magnitude is same for minimum and non-minimum phase T.F.s, assessment of relative stability sometime becomes erroneous.
(c) gain margin cannot be obtained for non-minimum phase systems.
(d) phase margin cannot be determined for non-minimum phase T.F.

32. The asymptotes of a Bode plots are

(a) the guiding straight lines of the magnitude plots.
(b) the guiding straight lines of the phase angle plot.
(c) magnitude plots are sketched using straight line segments and they are called asymptotes.
(d) the corrections made at corner frequencies are called asymptotes.

33. The corner frequency ω_c of a Bode plot is

(a) the frequency at which the frequency asymptote and high frequency asymptote meet.
(b) it is the reciprocal of the time constants of poles or zero.
(c) it is the frequency at which breakaway of magnitude occurs.
(d) it is the time constant of poles or zero.

34. What is the error introduced in a Bode magnitude plot at the breakaway frequency?

(a) +3 db for a pole.
(b) −3 db for a pole.
(c) −3 db for a zero.
(d) None of the above.

35. What is an octave?

(a) If the frequency is multiplied by a factor 10 frequency range is an octave.
(b) If the frequency is multiplied by a factor 8 frequency range is an octave.
(c) If the frequency is multiplied by a factor 6 frequency range is an octave.
(d) If the frequency is multiplied by a factor 2 frequency range is an octave.

36. What is a decade?

(a) If the frequency is multiplied by a factor 10, the frequency range is an decade.
(b) If the frequency is multiplied by a factor 8, the frequency range is an decade.
(c) If the frequency is multiplied by a factor 6, the frequency range is an decade.
(d) If the frequency is multiplied by a factor 4, the frequency range is an decade.

37. What is dB/octave?

(a) The slope of the straight line when the dB falls in the frequency range of one octave.
(b) The slope of the straight line when the dB raises in the frequency range of one octave.
(c) The slope of the straight line of the magnitude plot when the dB falls or raise in the frequency range of one octave is dB/octave.
(d) The slope of the phase plot when the dB falls or raise in the frequency range of one octave is called dB/octave.

38. What is dB/decade?

(a) The slope of the straight line when the dB falls in the frequency range of one decade is called dB/dec.
(b) The slope of the straight line when the dB raises in the frequency range of one decade is called dB/dec.
(c) The slope of the straight line of the magnitude plot when the dB falls or raises in the frequency range of one decade is dB/decade.
(d) The slope of the phase plot when the dB falls or raises in the frequency range of one decade is called dB/dec.

39. How dB/dec. and dB/oct. are related?

(a) 6 dB/oct = 20 dB/dec.
(b) 8 dB/oct = 20 dB/dec.
(c) 2 dB/oct = 20 dB/dec.
(d) 1 dB/oct = 20 dB/dec.

40. The shape of the straight line magnitude plot is 12 dB/dec. For the frequency range $\omega_1 = 5$ rad/sec. to $\omega_2 = 80$ rad/sec; the increase in dB is

(a) 96 dB.
(b) 48 dB.
(c) 192 dB.
(d) 24 dB.

41. Consider the system whose T.F. $G(s) = (s+4)^3$. **What is the corner frequency?**

(a) $\omega_c = 3\sqrt{4}$ rad/sec.
(b) $\omega_c = \sqrt{64}$ rad/sec.
(c) $\omega_c = 4$ rad/sec.
(d) $\omega_c = 64$ rad/sec.

42. For the example given in question (41) what is the error at corner frequency?

(a) $e_c = 3$ dB.
(b) $e_c = 6$ dB.
(c) $e_c = 9$ dB.
(d) $e_c = 27$ dB.

43. What is the magnitude and phase angle of the time delay whose T.F.?

$$G(s) = e^{-sT_d}$$

(a)

$$G(j\omega) = 1 \underline{\left| \omega T_d \right.}$$

(b)

$$G(j\omega) = 1 \underline{\left| -\omega T_d \times 57.3^\circ \right.}$$

(c)

$$G(j\omega) = 1 \underline{\left| -\omega T_d \pi \text{ rad} \right.}$$

(d)

$$G(j\omega) = 1 \underline{\left| \omega T_d \pi \text{ rad} \right.}$$

44. Sketch the polar plot for the following T.F. (Fig. 3.9)

$$G(s) = \frac{10}{s^2+4}$$

45. The polar plot for (Fig. 3.10)

$$G(s) = 10e^{-1s}$$

46. The Bode magnitude plot for the following T.F. is (Fig. 3.11)

$$G(s) = \frac{12}{(s+5)}$$

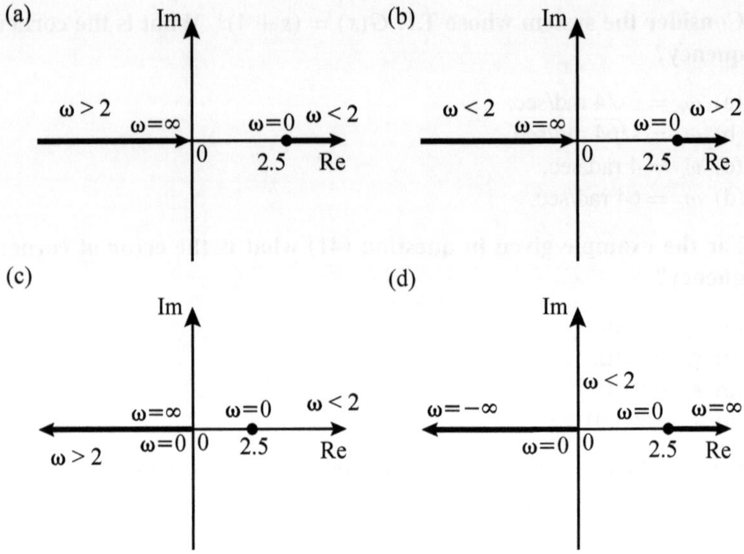

Fig. 3.9 Polar plot for $G(s) = \frac{10}{(s^2+4)}$

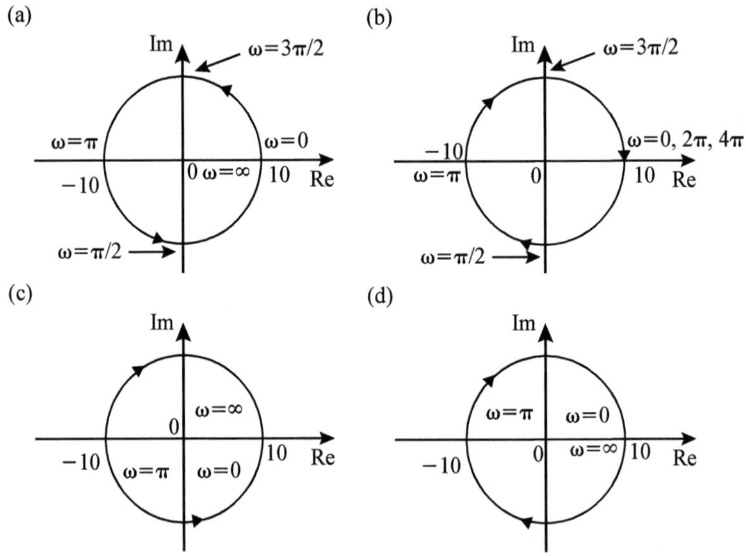

Fig. 3.10 Polar plot for $G(s) = 10e^{-1s}$

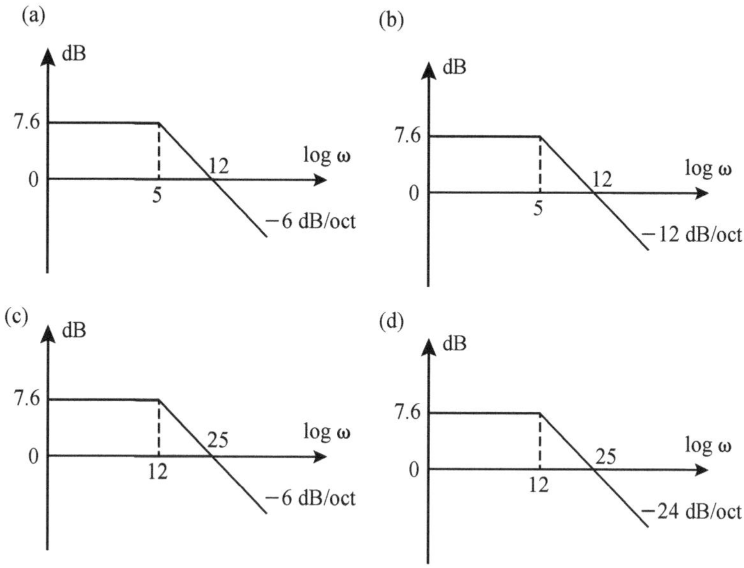

Fig. 3.11 Bode magnitude plot for $G(s) = \frac{12}{(s+5)}$

47. **Consider the following T.F.**

$$G(s) = \frac{24}{(1 + (s/5))}$$

sketch Bode magnitude plot from Bode plot the corner frequency ω_c **and gain cross over frequency** ω_{gc}. (Fig. 3.12)

48. **Consider the following T.F.**

$$G(s) = \frac{40}{s^2(1 + (s/5))}$$

Sketch the magnitude Bode plat and determine $omega_c$ **and** ω_{gc} (Fig. 3.13).

49. **What is the damping factor for 100% over-shot and what is the corresponding resonance peak?**

(a) $\zeta = 1;\ M_r = 1.$
(b) $\zeta = 0;\ M_r = \infty.$
(c) $\zeta = 0;\ M_r = 0.$
(d) $\zeta = 1;\ M_r = \infty.$

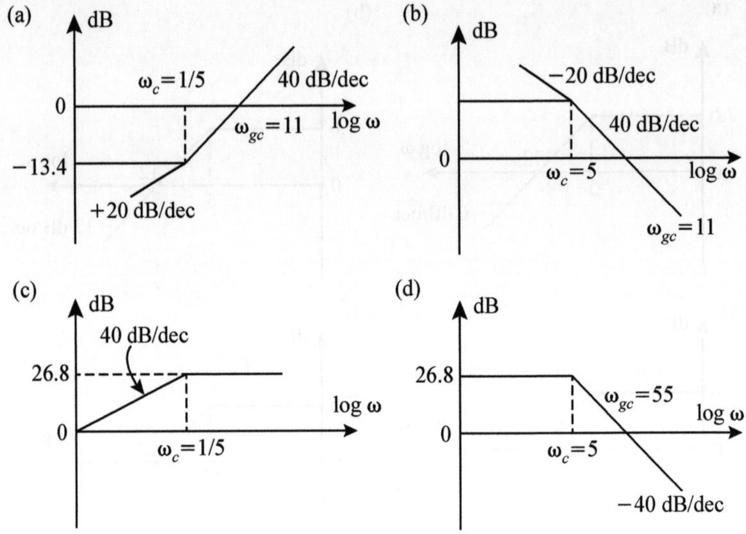

Fig. 3.12 Bode magnitude plot for $G(s) = \frac{24}{(1+(s/5))}$

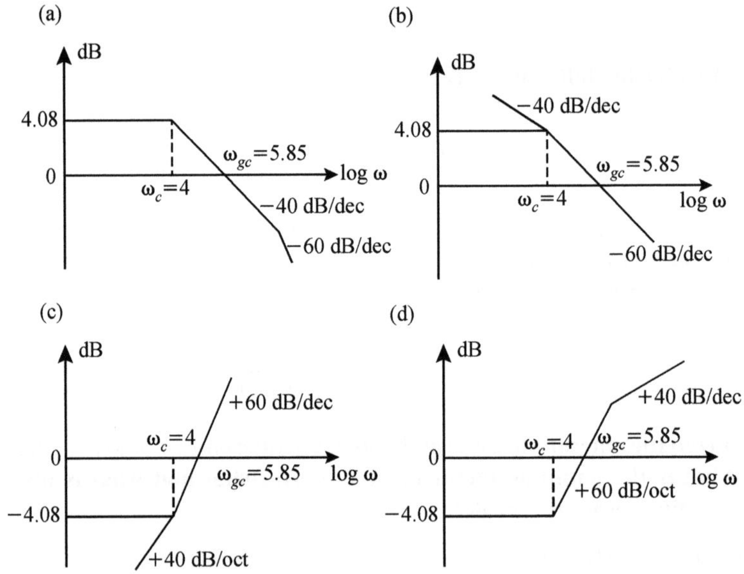

Fig. 3.13 Bode magnitude plot for $G(s) = \frac{40}{s^2(1+(s/5))}$

50. The magnitude phase plot shows that

$$G(j\omega)|_{\omega=5} = -18 \text{ dB} \underline{|180^\circ}$$

$$G(j\omega)|_{\omega=2} = 0 \text{ dB} \underline{|-140^\circ}$$

The gain margin and phase margin are

(a) $G_m = -18$ dB and $\phi_m = -40°$.
(b) $G_m = -18$ dB and $\phi_m = +40°$.
(c) $G_m = 18$ dB and $\phi_m = 40°$.
(d) $G_m = 18$ dB and $\phi_m = -45°$.

51. Generally the recommended values of phase margin and gain margins are

(a) $60° - 90°$; $G_m = 3$ dB.
(b) $30° - 60°$; $G_m = 3$ dB.
(c) $60° - 90°$; $G_m = 6$ dB.
(d) $30° - 90°$; $G_m = 6$ dB.

52. What is the desirable slope of the log magnitude plot at the gain cross over frequency for the system to be stable?

(a) zero dB/dec.
(b) +20 dB/dec.
(c) −20 dB/dec.
(d) −40 dB/dec.

53. For a prototype second order system as the damping ratio increases, the polar plot gets

(a) enlarged.
(b) shrinked.
(c) remains constant.
(d) restricted to the fourth quadrant.

54. The resonant peak of a standard prototype second order system is

$$M_r = \frac{1}{2\zeta\sqrt{1-\zeta^2}}$$

The recommended value for ζ is

(a) $0 \leq \zeta \leq 0.5$.
(b) $0 \leq \zeta \leq 0.707$.
(c) $0.5 \leq \zeta \leq 0.707$.
(d) $\zeta = 1.0$.

55. Large bandwidth means

(a) the system will pass wide range of frequency of the input signal.
(b) higher the bandwidth smaller is the rise time and the transient is faster.
(c) bandwidth is inversely proportional to rise time in the transient response.
(d) all the above.

56. The unique disadvantage of large bandwidth is

(a) for flat frequency response characteristic which gives higher bandwidth the cost of equipment is high for this.
(b) higher bandwidth allows noise to pass through the system.
(c) certain system requires low bandwidth.
(d) all the above.

57. A certain closed loop system has the following forward path T.F.

$$G(s) = \frac{64}{s(s+8)}$$

The bandwidth of the system is

(a) $\omega_b = 8$ rad/sec.
(b) $\omega_b = 10.18$ rad/sec.
(c) $\omega_b = 12.06$ rad/sec.
(d) $\omega_b = 13.15$ rad/sec.

58. The forward path T.F. of a certain unity feedback system is

$$G(s) = \frac{16}{s(s+8)}$$

The resonant peak and peak over-shoot are

(a) $M_r = 0.707$; $M_p = 14\%$.
(b) $M_r = 1.0$; $M_p = 0$.
(c) $M_r = 0.5$; $M_p = 18.5\%$.
(d) $M_r = 1.2$; $M_p = 0$.

59. The forward path T.F. of a certain unity feedback control system is given by

$$G(s) = \frac{K}{s(s+A)}$$

The desired resonant peak and resonant frequency are $M_r = 1.2$ **and** $\omega_r = 14$ **rad/sec. The values of** K **and** A **are**

(a) $K = 300.15$; $A = 20.5$.
(b) $K = 354.72$; $A = 17.82$.

(c) $K = 106.17;\ A = 12.84$.
(d) $K = 75.55;\ A = 8.26$.

60. The magnitude of frequency response of an under damped system is 5 at a frequency of 0 rad/sec and it peaks to $\sqrt{5/3}$ at a frequency of $\sqrt{5/3}$ rad/sec. The required T.F. is

(a)

$$M(s) = \frac{396}{(s^2 + 7.64s + 79)}.$$

(b)

$$M(s) = \frac{500}{(s^2 + 2.2s + 64)}.$$

(c)

$$M(s) = \frac{144}{(s^2 + 2s + 144)}.$$

(d)

$$M(s) = \frac{179.6}{(s^2 + 7.164s + 211.2)}.$$

61. A certain second order closed loop system when subjected to step input exhibits a peak over-shoot of 16.3% and the rise time is observed as 0.2 sec. The resonant peak is

(a) $M_r = 1.5$.
(b) $M_r = 0.707$.
(c) $M_r = 1.155$.
(d) $M_r = 1.676$.

62. A certain unity feedback second order system when subjected to sinusoidal input has a resonant peak of 1.2 and the corresponding frequency is 8 rad/sec. When the same system is subjected to step input, the percentage over-shoot is

(a) $M_p = 20.2\%$.
(b) $M_p = 18.52\%$.
(c) $M_p = 16.05\%$.
(d) $M_p = 10.22\%$.

63. A certain control system has the following T.F.

$$G(s) = \frac{(5s^2 + 2s + 10)}{(2s^2 - 3s - 2)}$$

The gain margin is

(a) $G_m = 5$ dB.
(b) $G_m = 7.62$ dB.
(c) $G_m = 1.732$ dB.
(d) $G_m = 3.52$ dB.

64. Consider the following T.F.

$$G(s) = \frac{100}{s(s+2)^2}$$

The phase margin and gain margin are

(a) $\phi_m = -40.6°$; $G_m = -0.16$ dB.
(b) $\phi_m = 40.6°$; $G_m = 0.16$ dB.
(c) $\phi_m = -20.5°$; $G_m = 1.26$ dB.
(d) $\phi_m = 15.5°$; $G_m = 6.25°$.

65. For a system to be stable, both phase margin and gain margin should be

(a) positive.
(b) negative.
(c) phase margin positive and gain margin negative.
(d) phase margin negative and gain margin positive.

66. The phase cross-over frequency is

(a) the frequency at which the phase angle of the T.F. $G(j\omega)$ is $180° + G(j\omega)$ at unity gain.
(b) the frequency at which the magnitude of the T.F of $G(j\omega)$ is unity.
(c) the frequency at which the phase angle of the T.F. $G(j\omega)$ is zero.
(d) the frequency at which the magnitude of $G(j\omega)$ becomes ∞.

67. The gain cross over frequency is the frequency at which

(a) $\angle G(j\omega) = 180°$
(b) $\angle G(j\omega) = 180°$
(c) $\angle G(j\omega) = +90°$
(d) $\angle G(j\omega) = 0°$

68. The phase margin in the polar plot is

(a) the phase angle of $G(j\omega)$ at unity gain.
(b) the phase angle of $G(j\omega) + 180°$ at unity gain.
(c) $\angle G(j\omega) + 180°$
(d) $\angle -G(j\omega) + 180°$

69. The gain margin in the polar plot is

(a) the reciprocal of $G(j\omega)|$.
(b) the reciprocal of $G(j\omega)|$ at $\angle G(j\omega)$ is 180°.
(c) the reciprocal of $G(j\omega)|$ at $\angle G(j\omega) = 90°$.
(d) the reciprocal of $G(j\omega)|$ at $\angle G(j\omega) = -90°$.

70. A linear minimum phase system is said to be absolutely stable

(a) if both the phase margin and gain margin are positive.
(b) if both the phase margin and gain margin are negative.
(c) if the phase margin is positive and gain margin is negative.
(d) if both phase margin and gain margin are positive and $\omega_{pc} > \omega_{gc}$.

71. A linear minimum phase system relative stability is higher if

(a) the gain margin is high and phase margin is low.
(b) the gain margin is low and phase margin is high.
(c) both the phase margin and gain margin are high.
(d) both the phase margin and gain margin are low.

72. Consider the following statements. Which one of them is wrong?

(a) To the phase margin is calculated at gain cross over.
(b) Gain margin is calculated and phase cross over.
(c) For the system to be stable pm and gm should be +ve and $\omega_{pc} > \omega_{gc}$.
(d) For the system to be stable $\omega_{pc} < \omega_{gc}$.

73. From the Bode plots the stability of the system is ensured if

(a) gain margin has $-$ve dB ω_{pc} and phase angle of $\angle G(j\omega_{gc}) < 108°$
(b) gain margin has $+$ve dB ω_{pc} and phase angle of $\angle G(j\omega_{gc}) < 108°$
(c) gain margin has $-$ve dB ω_{pc} and phase angle of $\angle G(j\omega_{gc}) > 108°$
(d) gain margin has $+$ve dB ω_{pc} and phase angle of $\angle G(j\omega_{gc}) > 108°$

74. Log magnitude–phase plot is a plot of

(a) $20 \log_{10} |G(j\omega)|$ versus $\angle G(j\omega)$ for $0 < \omega < \infty$.
(b) $\angle G(j\omega)$ versus ω for $0 < \omega < \infty$.
(c) $20 \log_{10} |G(j\omega)|$ versus ω for $0 < \omega < \infty$.
(d) $20 \log_{10} |G(j\omega)|$ versus $\log \omega$ for $0 < \omega < \infty$.

75. The closed loop frequency response can be more easily obtained using

(a) constant *M* circles.
(b) constant *N* circles.

(c) constant M and N circles.
(d) Nichol's chart.

76. System gain adjustment can be easily obtained using

(a) Polar plot.
(b) Bode plots.
(c) Log magnitude-phase plot.
(d) Constant M and N circles.

77. Nichol's chart is derived from

(a) polar plot.
(b) Bode plots.
(c) constant M circles.
(d) constant M and constant N circles.

78. The unique advantage of log magnitude versus phase plot is

(a) information about all frequency response specifications such as ϕ_m, G_m, ω_{gc}, ω_{pc}, M_r, ω_r and the effect of time delay can be easily determined.
(b) the required gain adjustment can be easily obtained.
(c) stability and relative stability can be easily obtained.
(d) unlike Bode plot, magnitude and phase plots are given in one graph.

79. The unique disadvantage of log magnitude–phase plot is

(a) unlike Bode plot, the effect of poles and zeros is not represented.
(b) correlation between time domain and frequency domain specifications are cannot be directly determined.
(c) the frequency response as a function of gain k is not indicated clearly.
(d) the magnitude of $|G(j\omega)|$ is expressed in dB which is $20\log_{10}|G(j\omega)$.

80. The polar plot of $G(j\omega)$ of a certain system passes through $(-1, 0)$ point. The phase margin, gain margin, ω_{gc}, ω_{pc} and the stability are

(a) $\phi_m = 0°$; $G_m = 0$ dB; $\omega_{pc} = \omega_{gc}$; the system is oscillatory.
(b) $\phi_m = 180°$; $G_m = 0$ dB; $\omega_{pc} = \omega_{gc}$; the system is oscillatory.
(c) $\phi_m = 0$; $G_m = 1$ dB; $\omega_{pc} > \omega_{gc}$; the system is stable.
(d) $\phi_m = 0$; $G_m = 1$ dB; $\omega_{pc} < \omega_{gc}$; the system is unstable.

81. Consider the following T.F.

$$G(s) = \frac{Ks^2}{(1+2s)(1+0.02s)}$$

For $\omega_{gc} = 5$ rad/sec, the value of K is

(a) $K = 0.1$.
(b) $K = 0.025$.

(c) $K = 0.057$.
(d) $K = 1.2$.

82. A certain system has the following T.F.

$$G(s) = \frac{0.1e^{-Ts}}{s(1 + 0.1s)(1 + 0.5s)}$$

Find T for $\phi_m = +24°$.

(a) $T = 10.73$
(b) $T = 8.25$
(c) $T = 5.52$
(d) $T = 20.0$

83. The phase margin in the Bode plot is obtained from

(a) $\phi = 180° + \underline{/G(j\omega)}$ when the gain is 0 dB.
(b) $\phi = 180° - \underline{/G(j\omega)}$ when the gain is 0 dB.
(c) $\phi = \underline{/G(j\omega)}$ when the gain is 0 dB.
(d) $\phi = -\underline{/G(j\omega)}$ when the gain is 0 dB.

84. The gain cross-over frequency ω_{gc} in the Bode plot is obtained from

(a) the frequency where $\phi = \underline{/G(j\omega)}$ is 190°.
(b) the frequency where the dB of the magnitude is 0 dB.
(c) the frequency at which the dB of the magnitude is -3 dB.
(d) the frequency at which the phase angle $\phi = \underline{/G(j\omega)} = 90°$.

85. The gain margin in Bode plot is obtained from

(a) the magnitude of $|G(j\omega)|$ expressed in dB where the phase angle $\underline{/G(j\omega)} = 90°$.
(b) the magnitude of $|G(j\omega)|$ expressed in dB where the phase angle $\underline{/G(j\omega)} = 180°$.
(c) the magnitude of $|G(j\omega)|$ expressed in dB at the phase angle $\underline{/G(j\omega)} = -90°$.
(d) the magnitude of $|G(j\omega)|$ expressed in dB at the phase angle $\underline{/G(j\omega)} = 45°$.

86. The phase cross over frequency in the Bode plot is obtained from

(a) the frequency at which $\angle G(j\omega) = 180^\circ$.
(b) the frequency at which $\angle G(j\omega) = 90^\circ$.
(c) the frequency at which $\angle G(j\omega) = 45^\circ$.
(d) the frequency at which the dB of $|G(j\omega)| = 0$.

87. What is phase margin in the log–magnitude versus phase plot?

(a) The dB of $|G(j\omega)|$ at which $\angle G(j\omega) = 45^\circ$.
(b) The dB of $|G(j\omega)|$ at which $\angle G(j\omega) = 90^\circ$.
(c) The dB of $|G(j\omega)|$ at which $\angle G(j\omega) = 135^\circ$.
(d) The dB of $|G(j\omega)|$ at which $\angle G(j\omega) = 180^\circ$.

88. The gain margin in log magnitude–phase plot is

(a) $G_m =$ dB of $|G(j\omega_{pc})|$ at $\angle G(j\omega) = 45^\circ$.
(b) $G_m =$ dB of $|G(j\omega_{pc})|$ at $\angle G(j\omega) = 90^\circ$.
(c) $G_m =$ dB of $|G(j\omega_{pc})|$ at $\angle G(j\omega) = 135^\circ$.
(d) $G_m =$ dB of $|G(j\omega_{pc})|$ at $\angle G(j\omega) = 180^\circ$.

89. From Bode and log magnitude–phase plot for the system to be stable.

(a) The phase margin is +ve.
(b) The gain margin is +ve.
(c) $\omega_{pc} > \omega_{gc}$.
(d) All the above.

90. Of the three frequency response plots namely log magnitude–phase plot, Bode-plot and polar plot which one is ideally suited for analysis and design purposes.

(a) Log magnitude–phase plot.
(b) Bode plot.
(c) Polar plot.
(d) All the three plots depending upon specific application.

91. The bandwidth (BW) of a second order prototype system is given by

(a)

$$\omega_b = \sqrt{1 - 2\zeta^2 + \sqrt{(4\zeta^4 - 4\zeta^2 + 2)}}.$$

(b)

$$\omega_b = \sqrt{1 + 2\zeta^2 + \sqrt{(4\zeta^4 - 4\zeta^2 + 2)}}.$$

(c)

$$\omega_b = \omega_n\sqrt{1 - 2\zeta^2 + \sqrt{(4\zeta^4 - 4\zeta^2 + 2)}}.$$

(d)

$$\omega_b = \omega_n\sqrt{1 + 2\zeta^2 + \sqrt{(4\zeta^4 - 4\zeta^2 + 2)}}.$$

92. The correlation between time response and frequency response of a prototype seconder system is

(a) M_p in time domain and M_r in frequency domain are expressed in terms of damping ratio ζ.
(b) t_p the time at which M_p occurs in time domain and resonant frequency ω_r in frequency domain are expressed in terms of ζ and ω_n.
(c) If t_p and M_p are K own, in the time domain, then ω_r and M_r specifications in the frequency domain can be obtained.
(d) None of the above.

93. The forward path T.F. of a certain unity negative feedback system is given by

$$G(s) = \frac{64}{s(s+8)}$$

The bandwidth of the system is

(a) $w_b = 5.234$ rad/sec.
(b) $w_b = 7.071$ rad/sec.
(c) $w_b = 8.255$ rad/sec.
(d) $w_b = 10.176$ rad/sec.

94. The forward path T.F. of a certain unity negative feedback system is given by

$$G(s) = \frac{16}{s(s+8)}$$

The bandwidth of the system is

(a) $w_b = 2.5737$ rad/sec.
(b) $w_b = 3.1364$ rad/sec.
(c) $w_b = 4.1414$ rad/sec.
(d) $w_b = 6.666$ rad/sec.

95. For the question 94, what are the resonant peak and resonant frequency?

(a) $M_r = 0.211$; $\omega_r = 0.55$ rad/sec.
(b) $M_r = 1.0$; $\omega_r = 0$ rad/sec.
(c) $M_r = 1.2$; $\omega_r = 0$ rad/sec.
(d) $M_r = 1.3$; $\omega_r = 1.11$ rad/sec.

96. A certain unity negative feedback control system has the following forward path T.F.

$$G(s) = \frac{K}{s(s+b)}$$

The system is to have $M_r = 1.2$ and $\omega_r = 14$ rad/sec. Find ζ and ω_n.

(a) $\zeta = 0.88$ and $\omega_n = 18.834$ rad/sec.
(b) $\zeta = 0.473$ and $\omega_n = 18.834$ rad/sec.
(c) $\zeta = 0.225$ and $\omega_n = 25.55$ rad/sec.
(d) $\zeta = 0.126$ and $\omega_n = 30.01$ rad/sec.

97. For the question 96, find the values of K and b.

(a) $K = 354.72$; $b = 17.817$.
(b) $K = 400.22$; $b = 15.754$.
(c) $K = 252.156$; $b = 22.711$.
(d) $K = 169.87$; $b = 327.416$.

98. For the question 96, find the BW.

(a) BW $\omega_b = 15.652$ rad/sec.
(b) BW $\omega_b = 10.135$ rad/sec.
(c) BW $\omega_b = 20.254$ rad/sec.
(d) BW $\omega_b = 25.673$ rad/sec.

99. A certain negative feedback system when subjected to sinusoidal input has a resonant peak $M_p = 1.2$ at a frequency of 8 rad/sec. Determine the B.W. of the system.

(a) B.W. $\omega_b = 8$ rad/sec.
(b) B.W. $\omega_b = 10$ rad/sec.
(c) B.W. $\omega_b = 12$ rad/sec.
(d) B.W. $\omega_b = 14$ rad/sec.

100. For question 99, what is the peak over-short M_p when subjected to step input?

(a) % $M_p = 6.322$.
(b) % $M_p = 18.52$.
(c) % $M_p = 24.56$.
(d) % $M_p = 30.16$.

101. Consider the following sinusoidal T.F.

$$G(j\omega) = \frac{5(j\omega + 2)}{(j\omega + 1)(-j\omega + 4)}$$

Determine the frequency at which the phase angle $\angle G(j\omega) = 0$

(a) $\omega = \sqrt{5}$.
(b) $\omega = \sqrt{3}$.
(c) $\omega = \sqrt{2}$.
(d) $\omega = \sqrt{0.5}$.

102. For the question 101, find the frequency at which the magnitude is

$$|G(\omega)| = \frac{5}{3}$$

(a) $\omega = 1.414$ rad/sec.
(b) $\omega = 2.441$ rad/sec.
(c) $\omega = 3.732$ rad/sec.
(d) $\omega = 3.966$ rad/sec.

103. Consider the following sinusoidal T.F.

$$G(j\omega) = \frac{(1 - j\omega 2)(1 + j\omega 5)}{(1 + j\omega 2)(1 - j\omega 5)}$$

At what frequency the maximum phase angle of $\angle G(j\omega)$ is achieved?

(a) $\omega_m = \sqrt{10}$ rad/sec.
(b) $\omega_m = \frac{1}{\sqrt{10}}$ rad/sec.
(c) $\omega_m = \sqrt{5}$ rad/sec.
(d) $\omega_m = \frac{1}{\sqrt{5}}$ rad/sec.

104. Consider the question 103. The maximum phase angle contribution of $\angle G(j\omega)$ is

(a) $\phi_m = 35.58°$.
(b) $\phi_m = 42.78°$.
(c) $\phi_m = 50.78°$.
(d) $\phi_m = 55.78°$.

105. In the log magnitude vs. phase plot the effect of time delay is

(a) to shift the $G(j\omega)$ plot vertically up.
(b) to shift the $G(j\omega)$ plot vertically down.
(c) to shift the $G(j\omega)$ plot horizontally towards right.
(d) to shift the $G(j\omega)$ plot horizontally towards left.

106. In the log magnitude vs. phase plot, by increasing the gain K

(a) the $G(j\omega)$ plot vertically shifted up.
(b) the $G(j\omega)$ plot vertically shifted down.
(c) the $G(j\omega)$ plot horizontally shifted towards right.
(d) the $G(j\omega)$ plot horizontally shifted towards left.

107. For a second order system the damping ratio ζ and the phase margin ϕ_m are roughly related by

(a) $\zeta = \pi\phi_m$.
(b) $\zeta = \frac{\pi}{180^\circ}\phi_m$.
(c) $\zeta = \frac{\pi}{360^\circ}\phi_m$.
(d) $\zeta = \frac{\pi}{\phi_m}$.

108. Consider the following T.F. of a certain control system

$$G(s) = \frac{1}{s(s+1)^2}$$

What is gain cross-over frequency?

(a) $\omega_{gc} = 0.523$ rad/sec.
(b) $\omega_{gc} = 0.68$ rad/sec.
(c) $\omega_{gc} = 0.707$ rad/sec.
(d) $\omega_{gc} = 0.931$ rad/sec.

109. Consider the system described in question 108. The phase cross over frequency is given by

(a) $\omega_{pc} = 2.0$ rad/sec.
(b) $\omega_{pc} = 1.5$ rad/sec.
(c) $\omega_{pc} = 1.2$ rad/sec.
(d) $\omega_{pc} = 1.0$ rad/sec.

110. Consider the system described in question 108. The phase margin is given by

(a) $\phi_m = 15^\circ$.
(b) $\phi_m = 20^\circ$.
(c) $\phi_m = 25^\circ$.
(d) $\phi_m = 30^\circ$.

111. Consider the system described in question 108. The gain margin is

(a) $G_m = 1$.
(b) $G_m = 2$.

(c) $G_m = 3$.
(d) $G_m = 4$.

112. A certain control system has the following open loop T.F.

$$G(s) = \frac{K}{s(1 + 0.2s)(1 + 0.05s)}$$

Find the value of K so that the gain margin is +25 dB.

(a) $K = 0.714$.
(b) $K = 1.125$.
(c) $K = 1.41$.
(d) $K = 1.5$.

113. For the question 112, find the value of K if the desired phase margin is $+65°$,

(a) $K = 1.78$.
(b) $K = 5.0$.
(c) $K = 2.57$.
(d) $K = 3.56$.

114. The centre of the constant M circle is for $M < 1$

(a) $c = \frac{M^2}{(1 - M^2)^2}; 0$.
(b) $c = \frac{M}{(1 - M^2)^2}; 0$.
(c) $c = \frac{(1 - M^2)}{M^2}; 0$.
(d) $c = \frac{(1 - M^2)}{M^2}; 0$.

115. The centre of the constant M circle for $M > 1$ is

(a) $c = \frac{M^2}{(M^2 - 1)}; 0$.
(b) $c = \frac{M^2}{(M^2 - 1)^2}; 0$.
(c) $c = \frac{(M^2 - 1)^2}{M^2}; 0$.
(d) $c = \frac{(M^2 - 1)^2}{M}; 0$.

116. Consider the constant M circle for $M = 1$. The centre is at

(a) $c = -1; 0$.
(b) $c = +1; 0$.

(c) $c = -\frac{1}{2}; 0.$
(d) $c = +\frac{1}{2}; 0.$

117. The centre of the constant N circles is

(a) $x = \dfrac{1}{2}; y = \frac{1}{2N}.$
(b) $x = 1; y = \frac{1}{2N}.$
(c) $x = -1; y = \frac{1}{2N}.$
(d) $x = -\dfrac{1}{2}; y = \frac{1}{2N}.$

118. The radius of the constant N circle is

(a)

$$r = \sqrt{\frac{1}{4} + \frac{1}{2N^2}}.$$

(b)

$$r = \sqrt{\frac{1}{4} + \frac{1}{4N^2}}.$$

(c)

$$r = \frac{1}{\sqrt{\frac{1}{4} + \frac{1}{2N^2}}}.$$

(d)

$$r = \frac{1}{\sqrt{\frac{1}{4} + \frac{1}{4N^2}}}.$$

119. The radius of the constant M circle for $M < 1$ is

(a)

$$r = \frac{M}{(1 + M^2)}.$$

(b)

$$r = \frac{(1 + M^2)}{M}.$$

(c)

$$r = \frac{M}{(1 - M^2)}.$$

(d)

$$r = \frac{M^2}{(1 - M^2)}.$$

120. The radius of the constant M circle for $M > 1$ is

(a)

$$r = \frac{M}{(1 + M^2)}.$$

(b)

$$r = \frac{M}{(M^2 - 1)}.$$

(c)

$$r = \frac{M^2}{(M^2 - 1)}.$$

(d)

$$r = \frac{M}{(M - 1)^2}.$$

Answers

(1) b	(13) a	(25) b	(37) c	(49) b
(2) c	(14) d	(26) a	(38) c	(50) c
(3) c	(15) b	(27) b	(39) a	(51) d
(4) a	(16) a	(28) c	(40) b	(52) c
(5) c	(17) c	(29) c	(41) c	(53) b
(6) b	(18) d	(30) c	(42) c	(54) b
(7) d	(19) a	(31) a	(43) b	(55) d
(8) c	(20) c	(32) c	(44) a	(56) d
(9) c	(21) b	(33) a	(45) b	(57) b
(10) b	(22) a	(34) b	(46) a	(58) b
(11) b	(23) a	(35) d	(47) b	(59) b
(12) c	(24) d	(36) a	(48) b	(60) a

(61) c
(62) b
(63) d
(64) a
(65) a
(66) a
(67) a
(68) b
(69) b
(70) d
(71) c
(72) d
(73) a
(74) a
(75) d
(76) c
(77) d
(78) a
(79) a
(80) a
(81) c
(82) a
(83) b
(84) b
(85) b
(86) a
(87) d
(88) d
(89) d
(90) a
(91) c
(92) c
(93) d
(94) a
(95) b
(96) b
(97) a
(98) d
(99) d
(100) b
(101) c
(102) a
(103) b
(104) c
(105) d
(106) a
(107) c
(108) b
(109) d
(110) b
(111) b
(112) c
(113) a
(114) a
(115) b
(116) c
(117) d
(118) b
(119) c
(120) b

Chapter 4
Stability Analysis of Linear System

Multiple Choice Questions

Choose the Correct Answer

1. **The unique disadvantage of system being unstable is**
   (a) for an unstable system the time response grows without bound.
   (b) the importance specifications are not satisfied.
   (c) time domain and frequency domain specifications have no meaning.
   (d) the time response grows without bound and causes severe damage not only to the system itself but also to the environment.

2. **What is internal or asymptotic stability?**
   (a) If the input is bounded the output is also bounded.
   (b) As time increases the output comes to steady level.
   (c) If the system is given small disturbance, the zero initial states go to non-zero initial states. After the disturbance if the non-zero initial states come to zero initial states the system is asymptotically stable.
   (d) Pole zero cancellation should not take place.

3. **What is external or BIBO stability?**
   (a) Zero input response should be bounded.
   (b) The zero state response should be bounded.
   (c) Zero input and zero out responses should be simultaneously bounded.
   (d) If for every bounded input, if the output is also bounded, the BIBO (external) stability is guaranteed.

© The Author(s), under exclusive license to Springer Nature Switzerland AG 2026

S. Palani, *Multiple-Choice Questions in Control Systems Engineering*,
https://doi.org/10.1007/978-3-031-93787-3_4

4. What is relative stability?

(a) If the system is stable and the gain margin is +ve and high the relative stability is high in the frequency domain.
(b) For stable system if the phase margin is high and +ve, the relative stability is high in the frequency domain.
(c) For stable system if the real part of the dominant roots is away from the $j\omega$ axis, the relative stability is high in the time domain.
(d) All the above.

5. How impulse response gives stability of the linear system?

(a) If the area of the impulse response is not finite the system is oscillatory.
(b) If the area under the impulse response curve is not finite the system is stable.
(c) If the area under the impulse response is finite the system is BIBO stable.
(d) None of the above.

6. How BIBO and asymptotic stability are correlated?

(a) BIBO stability ensures asymptotic stability.
(b) A symptotic stability does not guarantee BIBO stability.
(c) If the system is both BIBO and asymptotically stable the whole system is stable.
(d) Pole zero cancellation ensures the stability of the whole system.

7. Two systems with the following T.F. are connected in cascade. Is the system BIBO and asymptotically stable?

$$H_1(s) = \left(1 - \frac{4}{(s+2)}\right)$$

$$H_2(s) = \left(\frac{1}{(s-2)}\right)$$

Is the combined system asymptotically and BIBO stable?

(a) The system is BIBO stable.
(b) The system is not asymptotically unstable.
(c) Due to pole zero cancellation the system is both BIBO and asymptotically stable.
(d) The overall system is unstable.

8. A first order system has its poled placed in the −ve real axis.

(a) The system is BIBO stable since the area under the impulse response curve is finite.
(b) The system pole is in LHP which ensures asymptotic stability.
(c) The system is both BIBO and asymptotically stable.
(d) None of the above.

9. A system with a pure integrator only is

(a) asymptotically unstable.
(b) BIBO unstable.
(c) marginally stable.
(d) marginally stable but BIBO unstable.

10. A second order system has its complex conjugate poles in LHP. The system is

(a) marginally stable.
(b) asymptotically stable.
(c) BIBO stable.
(d) asymptotically and BIBO stable.

11. A certain second order system has its poles placed in the $j\omega$ axis. The system is

(a) the system is asymptotically unstable.
(b) the system is BIBO unstable.
(c) marginally stable.
(d) the system is marginally stable but asymptotically and BIBO unstable.

12. A certain first order system has its pole placed in RHP. The system is

(a) asymptotically stable.
(b) asymptotically unstable.
(c) BIBO stable.
(d) asymptotically and BIBO unstable.

13. A system has a double integrator. The system is

(a) asymptotically unstable.
(b) marginally unstable.
(c) BIBO unstable.
(d) marginally, BIBO and asymptotically unstable.

14. A second order system has its complex conjugate poles in RHP. The system is

(a) marginally unstable.
(b) BIBO unstable.
(c) asymptotically unstable.
(d) BIBO and asymptotically unstable.

15. A fourth order system has its poles repeated in the $j\omega$ axis. The system is

(a) BIBO marginally and asymptotically unstable.
(b) BIBO unstable.
(c) asymptotically unstable.
(d) marginally unstable.

16. BIBO stable system need not be asymptotically stable due to

(a) area under impulse response need not be zero.
(b) pole zero cancellation.
(c) presence of a pole at the origin of the s-plane.
(d) the presence of complex conjugate poles in the $j\omega$ axis.

17. For the following characteristic polynomial of a certain system find whether it is BIBO or asymptotic stable

$$F(s) = (s + 2)(s + 2 + j4)(s + 2 - j4)$$

(a) asymptotically stable.
(b) BIBO stable.
(c) marginally stable.
(d) BIBO and asymptotically stable.

18. The characteristic polynomial of a certain system is given by

$$F(s) = (s - 2)(s + 2 \pm j\omega 3)$$

Find the stability.

(a) asymptotically stable.
(b) asymptotically unstable.
(c) BIBO unstable.
(d) BIBO and asymptotically unstable.

19. The characteristic polynomial of a certain system is given by

$$F(s) = (s + 1)(s^2 + 9)$$

Find the stability.

(a) marginally stable.
(b) BIBO unstable.
(c) asymptotically unstable.
(d) BIBO and asymptotically unstable.

20. The characteristic polynomial of a certain system is given by

$$F(s) = (s + 1)(s^2 + 9)^2$$

Find the system stability.

(a) asymptotically stable.
(b) asymptotically unstable.

(c) BIBO unstable.
(d) BIBO and asymptotically unstable.

21. Consider the characteristic polynomial of a certain system which is given below

$$F(s) = s^2(s + 3)$$

The system is

(a) asymptotically unstable.
(b) BIBO unstable.
(c) BIBO and asymptotically unstable.
(d) marginally unstable.

22. The unique advantage of Routh-Hurwitz stability criterion is

(a) absolute stability of the system is determined.
(b) relative stability in the time domain is determined.
(c) it is possible to find the number of LHP, RHP and $j\omega$ axis roots.
(d) it is possible to find the amount of gain that can be increased without loss of stability.

23. What is the unique disadvantage of Routh-Hurwitz stability criterion?

(a) If the system is stable, Routh-Hurwitz criterion does not tell how stable it is.
(b) If the system is unstable, the criterion does not tell how to improve the system stability.
(c) The criterion is applicable only for real value of coefficients of characteristic polynomial. For complex coefficients it is not valid.
(d) In Routh-Hurwitz criterion $j\omega$ axis is considered as the stability boundary. The criterion cannot be applied to any other boundary in the complex plane.

24. The statement of Routh-Hurwitz criterion is

(a) the number of roots of the characteristic equation with positive real parts is equal to the number of changes in sign of first column of Routh array.
(b) the number of roots of the characteristic equation with negative real parts is equal to the number of changes in sign of first column of Routh array.
(c) the number of Os present in a complete row gives the number of RHP roots in the characteristic equation.
(d) the number of Os present in the first column of complete Routh array gives number of RHP roots of the characteristic equation.

25. When a first column becomes zero in the Routh array?

(a) It is completed by forming the reciprocal of the polynomial and using its coefficients.
(b) By using epsilon method.

(c) By multiplying by a factor $(s + 1)$.
(d) Any one of the three methods.

26. When a complete row is becoming zero, how to complete Routh's array?

(a) The auxiliary equation is formed and by differentiating this equation with respect to s, the coefficients of the row are substituted.
(b) By using epsilon method.
(c) By using the coefficient of the reciprocal of characteristic polynomial.
(d) By multiplying $F(s)$ by $(s + 1)$ and using these coefficients.

27. For certain system the inspection of the first column of Routh array shows that there is no sign change. The system

(a) is stable.
(b) is not stable.
(c) stability can be determined from the inspection of auxiliary polynomials and other polynomials.
(d) forms auxiliary polynomial only.

28. The open loop T.F. of a certain us its feedback control system is given by

$$G(s) = \frac{K(s+1)}{s^3 + As^2 + 2s + 1}$$

For the closed loop system to be stable

(a) $K = 4; A = 0.7$.
(b) $K = 2; A = 0.75$.
(c) $K = 8; A = 1.5$.
(d) $K = 10; A = 2.5$.

29. Consider the following characteristic polynomial.

$$F(s) = s^4 + 2s^3 + 6s^2 + Ks + 10$$

The real value of K so that the system is stable is

(a) $K < 12$.
(b) $K < 6$.
(c) $K < 2$.
(d) for all values of K the system is unstable.

30. Consider the following characteristic polynomial.

$$F(s) = s^4 + 3s^3 + 12s^2 + 16s + K$$

Find the range of K for which the system is stable.

(a) $0 < K < 25$.
(b) $0 < K < 35.55$.
(c) $0 < K < 55.55$.
(d) $0 < K < 320$.

31. **Consider the following characteristic polynomial.**

$$F(s) = s^3 + 12s^2 + 47s + 60$$

Test whether the system has a relative stability of 2 units.

(a) The system has a relative stability of 1 units.
(b) The system has a relative stability of 1.5 units.
(c) The system has a relative stability of 2 units.
(d) The system has a relative stability of 0 units.

32. **Consider a unity feedback system with the following open loop T.F.**

$$G(s) = \frac{10e^{-sT}}{(s+1)(s^2+2s+4)}$$

The value of T for which the system is stable is

(a) $0 < K < 5.36$.
(b) $0 < K < 2.35$.
(c) $0 < K < 0.133$.
(d) $1 < K < 10$.

33. **The property of auxiliary equation of an even polynomial is**

(a) the roots of the even polynomial are symmetric about the origin and real.
(b) they could be on the $j\omega$ axis as simple roots or repeated roots.
(c) the roots are quadratical (complex roots symmetrically placed with respect to the $j\omega$ axis).
(d) all the above three.

34. **The Routh-Hurwitz stability criterion is developed for the input**

(a) impulse input.
(b) step input.
(c) sinusoidal input.
(d) linear system stability is independent of input.

35. **The unique advantage of Nyquist criterion is**

(a) it gives information about absolute stability of the system.
(b) it is possible to get information about phase margin and gain margin.

(c) if the closed loop system is unstable it suggests methods to improve the stability of the closed loop system.
(d) all the three above.

36. Poles of open loop transfer function $G(s)$ are

(a) poles of the closed loop T.F.
(b) poles of the characteristic equation $1 + GH(s)$.
(c) roots of the characteristic equation.
(d) zeros of the characteristic equation.

37. Poles of the closed loop T.F. are

(a) poles of the open loop T.F.
(b) zeros of the characteristic equation $[1 + GH(s)]$.
(c) zeros of open loop T.F.
(d) none of the above.

38. Consider the following characteristic polynomial.

$$F(s) = s^2 + 4s - 10$$

The point $s = 2 + j4$ in the s-plane is mapped in the $F(s)$ plane as

(a) $F(s) = 14 + j12$.
(b) $F(s) = -14 + j12$.
(c) $F(s) = +14 + j12$.
(d) $F(s) = -14 - j12$.

39. The s plane contour which is enclosing a zero of $F(s)$ moved in the clockwise direction will encircle

(a) the origin of the $F(s)$ plane in the same clockwise direction once.
(b) the origin of the $F(s)$ plane in the counter clockwise direction once.
(c) the $(-1, 0)$ (some denoted as $(-1 + j0)$) point of the $F(s)$ plane with the same clockwise direction once.
(d) the $(-1, 0)$ point of the $F(s)$ plane with the counter clockwise direction once.

40. The s-plane contour which is enclosing a pole of $F(s)$ when moved in the clockwise direction will encircle

(a) the origin of the $F(s)$ plane in the clockwise direction once.
(b) the origin of the $F(s)$ plane in the counter clockwise direction once.
(c) $(-1, 0)$ point of the $F(s)$ plane in the counter clockwise direction once.
(d) $(-1, 0)$ point of the $F(s)$ plane in the clockwise direction once.

41. **There are 5 poles and 3 zeros of $F(s)$ inside the counter in the s-plane. This contour when moved in the clockwise direction will encircle the**

(a) origin of the $F(s)$ plane twice in the same clockwise direction.
(b) origin of the $F(s)$ plane twice in the counter clockwise direction.
(c) origin of the $F(s)$ plane 7 times in the same clockwise direction.
(d) origin of the $F(s)$ plane 7 times in the counter clockwise direction.

42. **There are 7 zeros and 4 poles of $F(s)$ inside the counter in the s-plane. There are 3 poles of $F(s)$ located outside this contour. The above s-plane contour when moved in the clockwise direction will encircle the origin of the $F(s)$ plane.**

(a) Will not at all encircle.
(b) 3 times in the same clockwise direction.
(c) 3 times in the counter clockwise direction.
(d) One encirclement in the clockwise direction.

43. **There are 4 poles and 2 zeros of $F(s)$ and they are enclosed by a contour. There are 2 zeros outside this contour. The above s-plane contour when moved in the clockwise direction will encircle the origin of the $F(s)$ plane.**

(a) Will not at all encircle.
(b) Two times in the same clockwise direction.
(c) Two times in the counter clockwise direction.
(d) Six times in the counter clockwise direction.

44. **The advantage of Nyquist stability criterion is**

(a) Nyquist stability criterion like Routh-Hurwitz criterion gives information about absolute stability of the system.
(b) Nyquist stability criterion gives information about relative stability in terms of phase margin and gain margin and ways of improving the stability margins.
(c) Since Nyquist criterion is developed in the frequency domain, the frequency domain specifications are easily obtained.
(d) Interpretation of Nyquist stability criterion is simple and straight forward. Further it requires open loop frequency response and open loop poles.

45. **According to Nyquist stability criterion, it is expressed by the following mathematical expression**

$$N_r = Z_r - P_r$$

where the symbols have the usual meaning. In the above equation P_r is

(a) P_r = number of open loop poles in RHP.
(b) P_r = number of closed loop system poles in RHP.
(c) P_r = number of roots of the characteristic equation in RHP.
(d) P_r = roots of the characteristic equation in RHP.

46. In the equation

$$N_r = Z_r - P_r$$

Z_r is

(a) zeros of the open loop T.F. in RHP.
(b) zeros of the closed loop system in RHP.
(c) zeros of the closed system in RHP.
(d) all of the above.

47. In the equation

$$N_r = Z_r - P_r$$

(a) N_r is the number of encirclements made by the $F(s)$ plane contour around the origin.
(b) N_r is the number of encircle made by the $F(s)$ plane contour around $(-1, 0)$ point.
(c) N_r is the number of rotation made by $G(j\omega)$ plot around $(-1, 0)$ in the $F(s)$ plane.
(d) N_r is the number of rotations made by $G(j\omega)$ plot around the origin in the $F(s)$ plane.

48. In the equation $N_r = Z_r - P_r$, when the system is stable? The system is made stable if

(a) $P_r = 0$.
(b) $Z_r = 0$.
(c) $N_r = 0$.
(d) all the above three cases.

49. What is the procedure to find the stability of the closed loop system to be stable.

(a) From the given $G(s)$, P_r is known. By Nyquist plot of $G(j\omega)$, the number of encirclement N_r around the $(-1, 0)$ point is determined. Using the above two, Z_r is found. Z_r should be 0 for the closed loop system to be stable.
(b) From $G(s)$, P_r is known. Z_r is known from the plot $G(j\omega)$ in the $GH(s)$ plane. Using these two N_r can be found. For a closed loop system to be stable $N_r = 0$.
(c) From $G(s)$, P_r is known. By $G(j\omega)$ plot N_r can be determined. Knowing these two P_r can be found. For a closed loop system to be stable $P_r = 0$.
(d) From $G(s)$, P_r is known. By $G(j\omega)$ plot Z_r is known. Knowing these two P_r can be found. For a closed loop system to be stable $N_r = 0$.

50. If the open loop system is stable, then according to Nyquist stability criterion, the closed loop system is also stable iff

(a) the s-plane Nyquist contour when mapped into $GH(s)$ plane, should not encircle $(-1, 0)$ point.
(b) the s-plane Nyquist contour when mapped into $GH(s)$ plane, should not encircle $(-1, 0)$ point in the same clockwise direction more than once.
(c) the s-plane Nyquist contour when mapped into $GH(s)$ plane, should not encircle $(-1, 0)$ point in the counter clockwise direction more than once.
(d) the s-plane Nyquist contour when mapped into $GH(s)$ plane it should encircle the origin of $G(s)$ plane once in the count clockwise direction.

51. If the open loop system is unstable, then for the closed loop system to be stable when this is possible according to Nyquist criterion?

(a) If the open loop system is unstable $(Pr \neq 0)$, then for the closed loop system to be stable $(Z_r = 0)$, the mapping Nyquist contour in the s-plane should not encircle $(-1, 0)$ point.
(b) If the open loop system is unstable $(Pr \neq 0)$, then for the closed loop system to be stable, the mapping Nyquist contour from s-plane to $GH(s)$ phase should encircle $(-1, 0)$ point as many times as there are number of open loop poles in RHS and the encirclement should be in the counter clockwise direction.
(c) If the open loop system is unstable then for the closed loop system to be stable, the mapping s-plane Nyquist contour to $GH(s)$ phase should encircle $(-1, 0)$ point in the $GH(s)$ and the encirclement should be in the same clockwise direction and the number of encirclement should be as many time s as these are open loop poles in RHP.
(d) All the above.

52. A certain control system has the following T.F. in the forward path

$$G(s) = \frac{K}{s(s + T_1)(s + T_2)}$$

Using Nyquist stability criterion establishes the relationship between the parameters so that the closed loop system is stable. What is phase cross over frequency?

(a)

$$\frac{K}{T_1 T_2 (T_1 + T_2)} < 1 \text{ the system is stable.}$$

and $\omega_{pc} = \sqrt{T_1 T_2}$.

(b)

$$\frac{K}{T_1T_2(T_1+T_2)} > 1 \text{ the system is unstable.}$$

and $\omega_{pc} = \sqrt{T_1T_2}$.

(c)

$$\frac{K}{T_1T_2(T_1+T_2)} = 1 \text{ the system is marginally stable.}$$

(d) All are correct and $\omega_c = \sqrt{T_1T_2}$.

53. Consider the following open loop T.F. of an unity feedback control system

$$G(s) = \frac{K}{(s+1)(s^2+2s+5)}$$

Find the gain margin phase cross over frequency. Find stability for $K = 10$.

(a) $G_m = \frac{16}{K}$; $\omega_{pc} = \sqrt{7}$; $Z_r = 2$ (unstable).
(b) $G_m = \frac{5}{K}$; $\omega_{pc} = \sqrt{10}$; $Z_r = 0$ (stable).
(c) $G_m = \frac{K}{16}$; $\omega_{pc} = \sqrt{10}$; $Z_r = 1$ (unstable).
(d) $G_m = K$; $\omega_{pc} = 5.0$; $Z_r = 0$ (stable).

54. A certain unity feedback system has the following open loop T.F.

$$G(s) = \frac{K(s+4)}{(s+1)(s-1)}$$

The value of K for the closed loop system to be stable is

(a) $K < 0.25$.
(b) $K = 0.25$.
(c) $K > 0.25$.
(d) $K = 0.1$.

55. A certain feedback control system has the following open loop T.F.

$$G(s) = \frac{s(1+4s)}{(s^3+2s+4)}$$

Find the gain margin ω_{pc} and the stability of the closed loop system.

(a) $G_m = 0.5$; $\omega_{pc} = 2$ rad/sec; $Z_r = 2$, the system is unstable.
(b) $G_m = 0.75$; $\omega_{pc} = 2.14$ rad/sec; $Z_r = 2$, the system is unstable.
(c) $G_m = 0.414$; $\omega_{pc} = 1.55$ rad/sec; $Z_r = 2$, the system is unstable.
(d) $G_m = 0.24$; $\omega_{pc} = 3.2$; $Z_r = 0$, the system is stable.

56. A certain unity feedback system has the following forward path T.F.

$$G(s) = \frac{K(s+3)(s+10)}{(s-4)(s-5)}$$

Find the range of K for which the closed loop system is stable. Find ω_{pc}.

(a) $\omega_{pc} = 4.91$ rad/sec; $K < 0.694$, the system is unstable.
(b) $\omega_{pc} = 4.91$ rad/sec; $K = 0.694$, the system is marginally unstable.
(c) $\omega_{pc} = 4.91$ rad/sec; $K > 0.694$, the system is stable.
(d) For all values of K, the system is unstable.

57. Consider the open loop T.F. of a certain unity feedback control system.

$$G(s) = \frac{K(s+0.5)(s+2)}{(1+0.2s)(1-5s)}$$

Find ω_{pc}. Find the range of K for which the system is stable

$$\omega_{pc} = 1.7815 \text{ rad/sec}$$

(a) $K > 1.92$ system is stable.
(b) $K = 1.92$ system is stable.
(c) $K < 1.92$ system is stable.
(d) for all values of K, the system is stable.

58. Consider the following T.F. of a certain unity feedback system

$$G(s) = \frac{K(s-0.5)}{(1+2s)^2}$$

Find the range of K for which the closed loop system is stable.

(a) $K > 0.2$ the system is stable.
(b) $K = 0.2$ the system is stable.
(c) $K < 0.2$ the system is stable.
(d) For all values of K, the system is unstable.

59. For a certain system, the frequency response reveals that $G_m = -6$ dB; $\omega_{pc} = 2$ rad/sec; $\omega_{gc} = 3$ rad/sec. Is the system stable?

(a) The system is stable.
(b) The system is marginally stable.
(c) The system is asymptotically stable.
(d) The system is unstable.

60. A certain unity feedback control system has the following forward path T.F.

$$G(s) = \frac{K}{(s+5)^3}$$

Find ω_{pc} and the range value of K for which the closed loop system is stable

$$\omega_{pc} = 8.66 \text{ rad/sec.}$$

(a) $K < 1000$ the system is stable.
(b) $K = 1000$ the system is stable.
(c) $K > 1000$ the system is stable.
(d) for all values of a K the system is unstable.

61. The forward path T.F. of a certain unity feedback system is

$$G(s) = \frac{1+Ks}{s(s-4)}$$

Find ω_{pc}. Find the range value of K for which the system is stable.

(a) $K > 4$; $\omega_{pc} = \frac{2}{\sqrt{K}}$; the system is stable.
(b) $K = 4$; $\omega_{pc} = \frac{1}{\sqrt{K}}$; the system is stable.
(c) $K < 4$; $\omega_{pc} = \frac{1}{K}$; the system is stable.
(d) $K > 2$; $\omega_{pc} = K$; the system is stable.

62. The forward path T.F. of a certain unity feedback control system is given below. For the closed loop system to be stable find the range of K and ω_{pc}.

$$G(s) = \frac{(1+Ks^2)}{s^3}$$

(a) $K < 0.8$; $\omega_{pc} = \frac{1}{K}$; the system is stable.
(b) $K = 0.8$; $\omega_{pc} = K$; the system is stable.
(c) $K > 0.8$; $\omega_{pc} = \frac{1}{K}$; the system is stable.
(d) the system is unstable for all values of K.

63. The loop T.F. of a certain closed loop system is

$$GH(s) = \frac{Ks(s+2)}{(s^3+64)}$$

Find the value of K and ω_{pc} so that the closed loop system is stable.

(a) $K > 5.66$; $\omega_{pc} = 3.36$ rad/sec, the system is stable.
(b) $K < 5.66$; $\omega_{pc} = 3.36$ the system is stable.

(c) $K = 5.66$; $\omega_{pc} = 3.36$ the system is stable.
(d) the system is unstable for all values of K.

64. A certain feedback control system has the following loop T.F.

$$GH(s) = \frac{K}{(s^2 + 3s + 10)(s^2 + 3s + 100)}$$

Find the range of K values and ω_{pc} for which the closed loop system is stable.

(a) $K > 2520$; $\omega_{pc} = 7.416$, the system is stable.
(b) $K < 2520$; $\omega_{pc} = 7.416$, the system is stable.
(c) $K = 2520$; $\omega_{pc} = 7.416$, the system is stable.
(d) For all values of K the system is stable.

65. Consider the loop T.F. of a certain feedback control system

$$GH(s) = \frac{K}{(1 + 2s)^3}$$

Find the value of K and ω_{pc} for which the closed loop system is stable.

(a) $K < 8$; $\omega_{pc} = 0.866$, the system is stable.
(b) $K > 8$; $\omega_{pc} = 0.866$, the system is stable.
(c) $K = 8$; $\omega_{pc} = 0.866$, the system is stable.
(d) the system is stable for all values of K.

66. The loop T.F. of a feedback control system is

$$GH(s) = \frac{K}{(1 + 2s)^4}$$

Find the value of K and ω_{pc} for the closed loop system is to be stable.

(a) for $K > 8$ and $\omega_{pc} = 0.5$, the system is stable.
(b) for $K = 8$ and $\omega_{pc} = 0.5$, the system is stable.
(c) for $K < 8$ and $\omega_{pc} = 0.5$, the system is stable.
(d) for all values of K, the system is stable.

67. Consider the following characteristic polynomial of a certain control system

$$F(s) = s^6 + 2s^5 + 8s^4 + 12s^3 + 20s^2 + 16s + 16$$

Find the stability of the system.

(a) stable.
(b) unstable.

(c) marginally stable.
(d) asymptotically stable.

68. Consider the following characteristic polynomial of a certain control system.

$$F(s) = s^6 + 4s^5 + 11s^4 + 12s^3 + 40s^2 + 64s + 48$$

Investigate the stability of the system.

(a) Stable.
(b) Unstable.
(c) Marginally stable.
(d) Asymptotically stable.

69. Consider the following open loop T.F. of a certain closed system

$$G(s) = \frac{K(s-2)^2}{(s+2)^2(s+4)^2}$$

Find the range of K for the closed loop system to be stable.

(a) $K < 8$ the system is stable.
(b) $K > 8$ the system is stable.
(c) $K = 8$ the system is stable.
(d) for all values of K the system is stable.

70. Consider closed loop system has the following open loop T.F.

$$GH(s) = \frac{Ks^3}{(s+2)^2(s+3)^2}$$

Find the range of K and ω_{pc} for the closed system is stable

$$\omega_{pc} = 1$$

(a) $K > 50$ the system is stable.
(b) $K < 50$ the system is stable.
(c) $K = 50$ the system is stable.
(d) For all values of K the system is unstable.

71. The $G(j\omega)$ plot of certain closed loop negative feedback system is shown in Fig. 4.1.The open system has one pole in RHP. Find whether the closed loop system is stable or unstable. Find the number of closed loop poles in RHP.

(a) $z_r = 1$; the closed loop system is unstable.
(b) $z_r = 2$; the closed loop system is unstable.

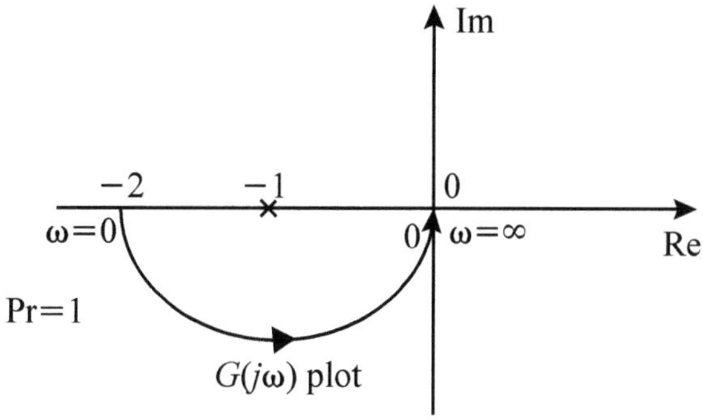

Fig. 4.1 Plot of $G(j\omega)$ for Example 71

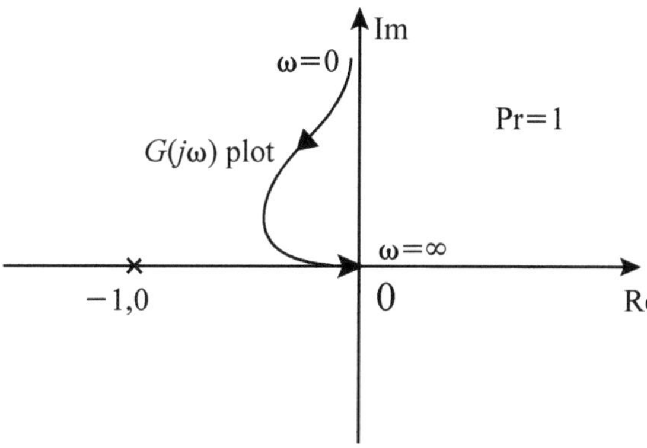

Fig. 4.2 $G(j\omega)$ plot for Question 72

(c) $z_r = 0$; the closed loop system is stable.
(d) $z_r = 1$; the closed loop system is stable.

72. The $G(j\omega)$ plot of certain closed loop system with negative feedback system is shown in Fig. 4.2. The open system has one pole in RHP. Determine the stability of the closed loop system and the number of closed loop poles in RHP.

(a) $z_r = 0$; the closed loop system is stable.
(b) $z_r = 1$; the closed loop system is unstable.
(c) $z_r = 2$; the closed loop system is unstable.
(d) $z_r = 0$; the closed loop system is unstable.

73. The $G(j\omega)$ plot of certain negative feedback system is shown in Fig. 4.3. The open system has one pole in RHP. Find the closed loop system stability and the number of closed loop poles in RHP.

(a) $z_r = 0$; the closed loop system is stable.
(b) $z_r = 1$; the closed loop system is unstable.
(c) $z_r = 2$; the closed loop system is unstable.
(d) $z_r = 3$; the closed loop system is unstable.

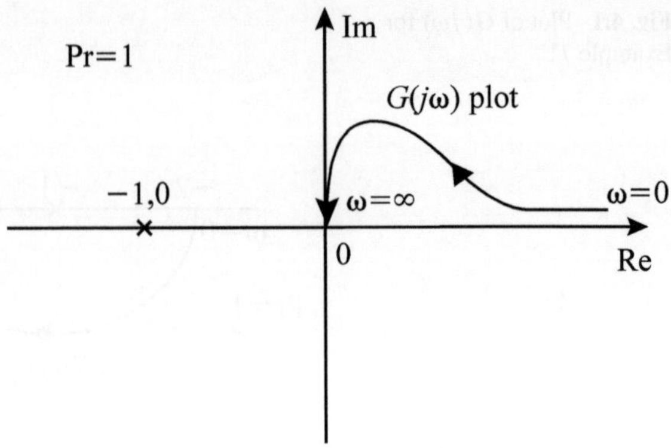

Fig. 4.3 $G(j\omega)$ plot for Question 73

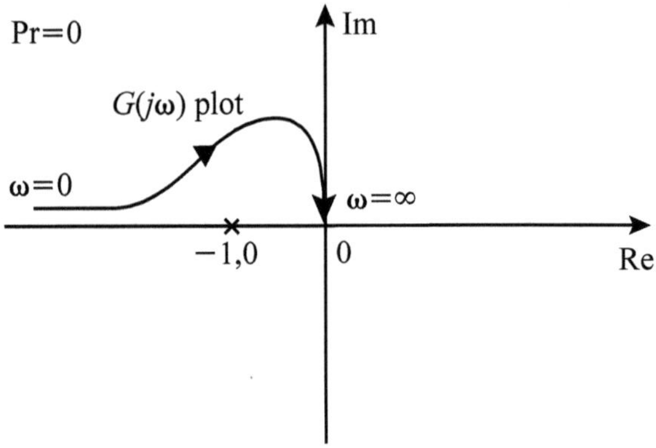

Fig. 4.4 $G(j\omega)$ for plot Question 74

74. The $G(j\omega)$ of a certain unity negative feedback minimum phase T.F. system is shown in Fig. 4.4. Determine whether closed loop system is stable and the number of closed loop poles in RHP.

(a) $z_r = 0$; the closed loop system is stable.
(b) $z_r = 1$; the closed loop system is unstable.
(c) $z_r = 2$; the closed loop system is unstable.
(d) $z_r = 0$; the closed loop system is unstable.

75. The $G(j\omega)$ plot of a certain closed loop minimum phase system is shown in Fig. 4.5. Find the value of the gain K so that the closed loop system is asymptotically stable.

(a) $K = 16$.
(b) $K > 16$.
(c) $K < 16$.
(d) The system is unstable for all values of K.

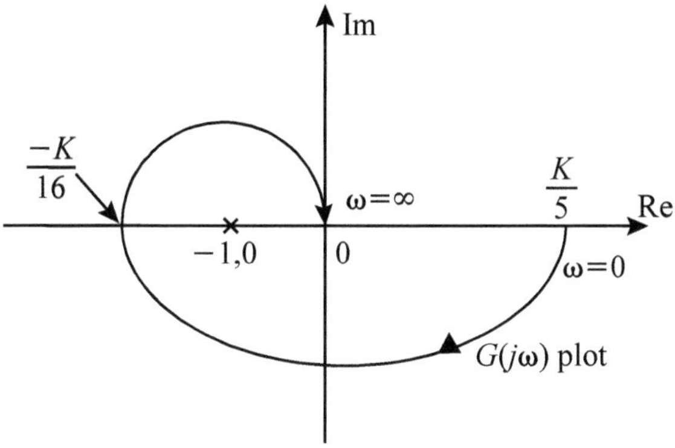

Fig. 4.5 $G(j\omega)$ plot for Question 75

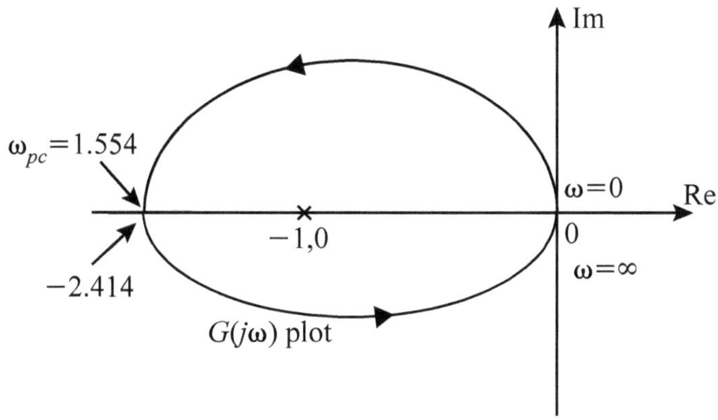

Fig. 4.6 $G(j\omega)$ plot for Question 76

76. The $G(j\omega)$ plot of a unity feedback minimum phase system is shown in Fig. 4.6. There are two open loop poles in RHP. Find whether the closed loop system is asymptotically stable.

(a) $z_r = 0$; the closed loop system is stable.
(b) $z_r = 1$; the closed loop system is unstable.
(c) $z_r = 2$; the closed loop system is stable.
(d) $z_r = 3$; the closed loop system is unstable.

77. The $G(j\omega)$ plot of a certain non-minimum phase proper open loop T.F. is shown in Fig. 4.7. The open loop system is unstable. Find z_r and the stability of the closed loop system.

(a) $z_r = 0$; the closed loop system is stable.
(b) $z_r = 1$; the closed loop system is unstable.

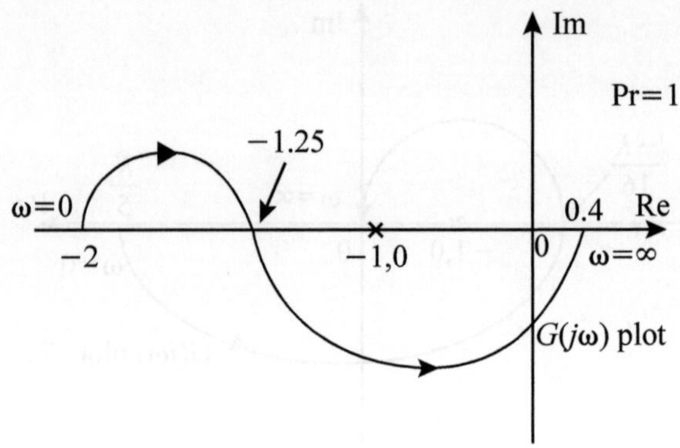

Fig. 4.7 $G(j\omega)$ plot for Question 77

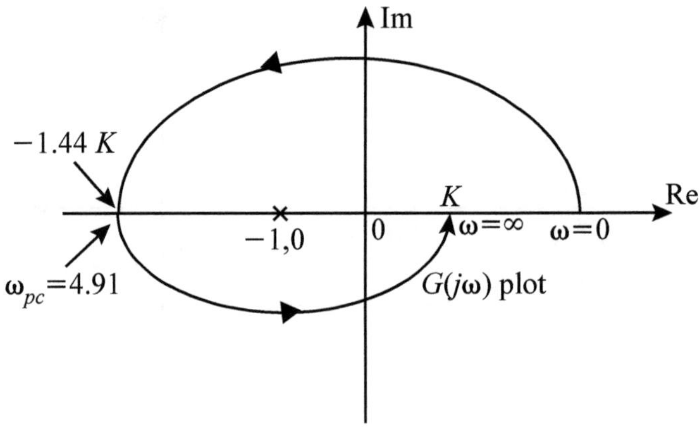

Fig. 4.8 $G(j\omega)$ plot for Question 78

(c) $z_r = 2$; the closed loop system is unstable.
(d) $z_r = 0$; the closed loop system is unstable.

78. The $G(j\omega)$ plot of a closed loop system with proper T.F. and two poles in RHS is shown in Fig. 4.8. Find the value of the gain K so that the closed loop system asymptotically stable.

(a) $K < 0.6946$.
(b) $K = 0.6046$.
(c) $K > 0.69416$.
(d) $0.694 < K < 1$.

79. The $G(j\omega)$ plot of a certain closed loop system is shown in Fig. 4.9. The open loop system T.F. is strictly proper with non-minimum phase and unstable with Pr= 1. Determine the value of K so that the closed loop system is asymptotically stable.

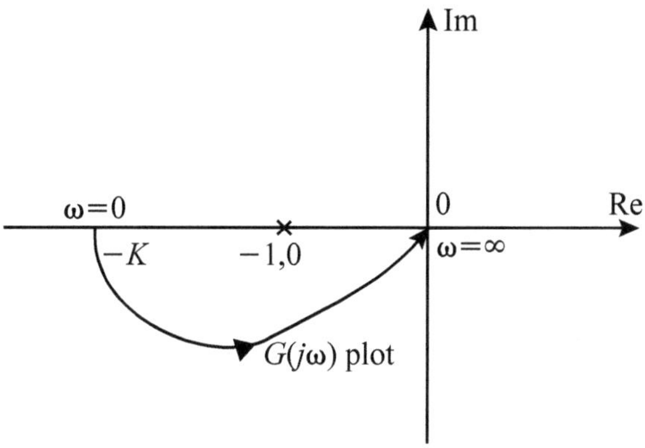

Fig. 4.9 $G(j\omega)$ plot for Question 79

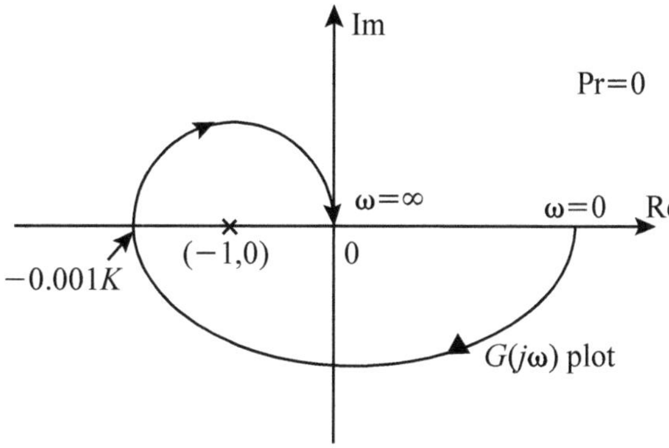

Fig. 4.10 $G(j\omega)$ plot for Question 80

(a) $K < 1$.
(b) $K = 1$.
(c) $K > 1$.
(d) $1 < K < 0$.

80. The $G(j\omega)$ plot of a strictly proper T.F. is shown in Fig. 4.10. Find the value of K so that the closed loop system is asymptotically stable.

(a) $K > 1000$.
(b) $K = 1000$.
(c) $K < 1000$.
(d) $1000 < K < 0$.

81. The $G(j\omega)$ plot of a strictly proper minimum phase open loop system with Pr$=0$ is shown in Fig. 4.11. Find the value of K so that the closed loop system is asymptotically stable.

(a) $K < 9.1 \times 10^{-3}$.
(b) $K = 9.1 \times 10^{-3}$.

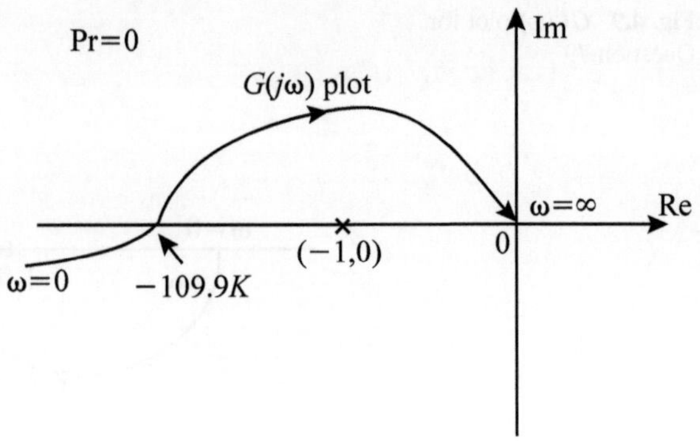

Fig. 4.11 $G(j\omega)$ plot for Question 81

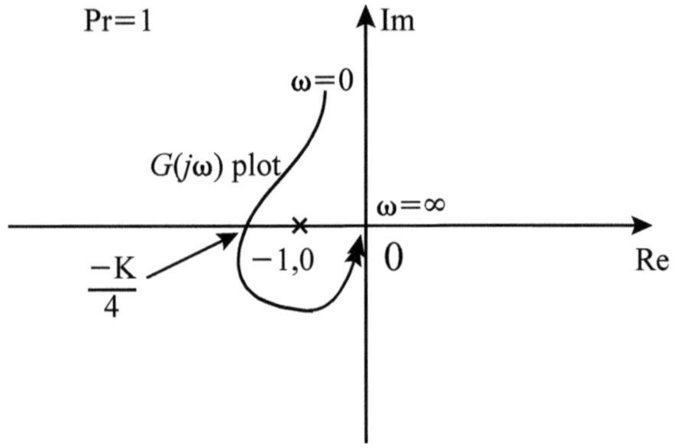

Fig. 4.12 $G(j\omega)$ plot for Question 82

(c) $K > 9.1 \times 10^{-3}$.
(d) $9.1 \times 10^{-3} < K < 0$.

82. The $G(j\omega)$ plot of a strictly proper non-minimum phase open loop system with Pr= 1 is shown in Fig. 4.12. Find the value of K so that the closed loop system is asymptotically stable.

(a) $K < 4$; the closed loop system is stable.
(b) $K = 4$; the closed loop system is stable.
(c) $K < 4$; the closed loop system is stable.
(d) $-4 < K < 4$; the closed loop system is stable.

83. The $G(j\omega)$ plot of a minimum phase with strictly proper T.F. of a open loop system is shown in Fig. 4.13. Find the value of K so that the closed loop system is asymptotically stable.

(a) $K > 5.66$; the closed loop system is stable.
(b) $K = 5.66$; the closed loop system is stable.
(c) $K < 5.66$; the closed loop system is stable.
(d) For all values of K the system is unstable.

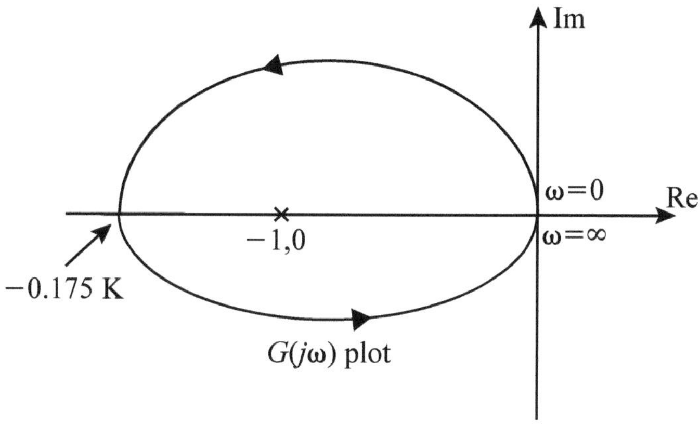

Fig. 4.13 $G(j\omega)$ plot for Question 83

Fig. 4.14 $G(j\omega)$ plot for Question 84

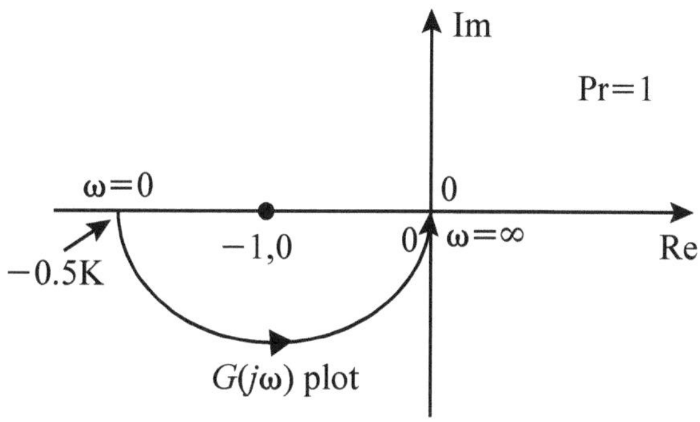

84. **The $G(j\omega)$ plot of a strictly proper T.F. of a open loop system is shown in Fig. 4.14. The open system has one RHP pole. Find the value of K so that the closed loop system is asymptotically stable.**

(a) $K < 2$; the closed loop system is stable.
(b) $K = 2$; the closed loop system is stable.
(c) $K > 2$; the closed loop system is stable.
(d) For all values of K the system is unstable.

85. **The $G(j\omega)$ plot of an open loop system with proper non-minimum phase T.F. is shown in Fig. 4.15. Find the value of K so that the closed loop system is asymptotically stable.**

(a) $K < 0.1$; the closed loop system is stable.
(b) $K = 0.1$; the closed loop system is stable.
(c) $K < 0.1$; the closed loop system is stable.
(d) For all values of K the closed loop system is unstable.

Fig. 4.15 $G(j\omega)$ plot for Question 85

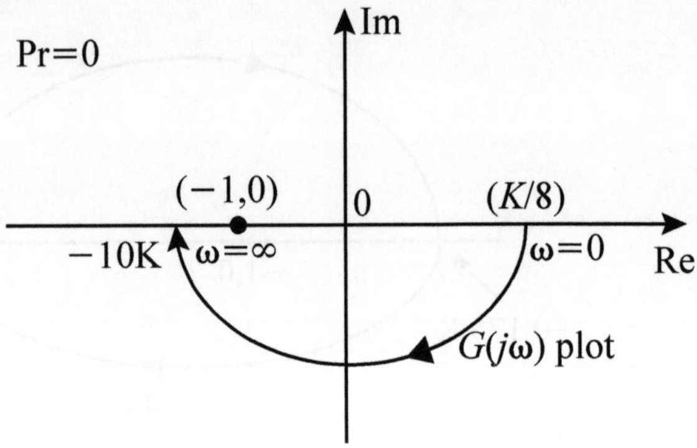

Fig. 4.16 $G(j\omega)$ plot for Question 86

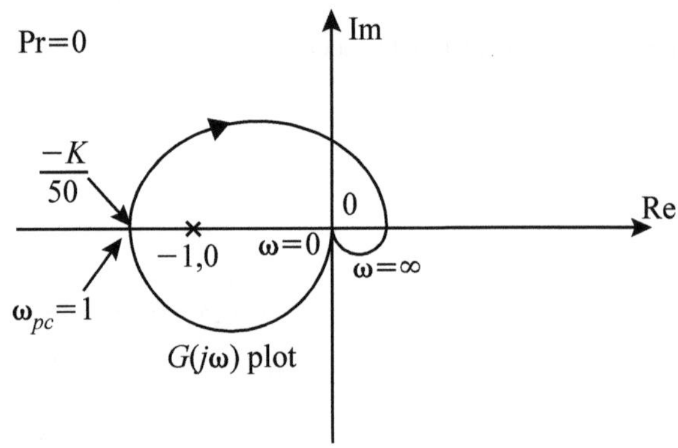

86. The $G(j\omega)$ plot of a unity negative feedback control system is shown in Fig. 4.16. The loop T.F. is strictly proper and minimum phase. Determine the value of K so that the closed loop system is asymptotically stable.

(a) $K < 50$; the closed loop system is stable.
(b) $K = 50$; the closed loop system is stable.
(c) $K < 50$; the closed loop system is stable.
(d) For all values of K the closed loop system is unstable.

87. The $G(j\omega)$ plot of a certain unity negative feedback control system is shown in Fig. 4.17. The system is minimum phase and strictly proper. Determine the value of K so that the closed loop system is asymptotically stable.

(a) $K < 4$; the closed loop system is stable.
(b) $K = 4$; the closed loop system is stable.
(c) $K > 4$; the closed loop system is stable.
(d) For all values of K the closed loop system is unstable.

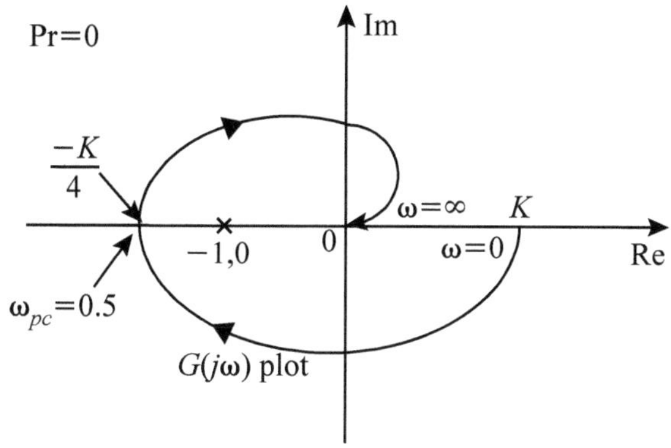

Fig. 4.17 $G(j\omega)$ plot for Question 87

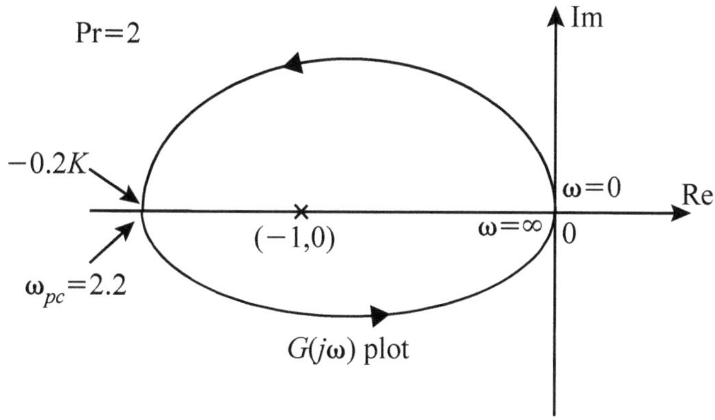

Fig. 4.18 $G(j\omega)$ plot for Question 88

88. The $G(j\omega)$ plot of a certain unity feedback is shown in Fig. 4.18. Find the value of K so that the closed loop system is asymptotically stable.

(a) $K > 5$; the closed loop system is stable.
(b) $K = 5$; the closed loop system is stable.
(c) $K < 5$; the closed loop system is stable.
(d) For all values of K the closed loop system is unstable.

89. A certain unity feedback control system has the following forward path T.F.

$$G(s) = \frac{K}{s(s + T_1)(s + T_2)}$$

The phase cross-over frequency of the frequency response plot is

(a)

$$\omega_{pc} = \sqrt{\frac{T_1}{T_2}}.$$

(b)

$$\omega_{pc} = \sqrt{\frac{T_2}{T_1}}.$$

(c)

$$\omega_{pc} = \sqrt{\frac{1}{T_1 T_2}}.$$

(d)

$$\omega_{pc} = \sqrt{T_1 T_2}.$$

90. For the system described in question 89, the limiting value of K, T_1 and T_2 for the closed loop system to be stable is

(a)

$$K\frac{(T_1 + T_2)}{T_1 T_2} > 1.$$

(b)

$$K\frac{T_1 T_2}{(T_1 + T_2)} > 1.$$

(c)

$$\frac{K}{T_1 T_2 (T_1 + T_2)} > 1.$$

(d)

$$\frac{K}{T_1 T_2 (T_1 + T_2)} < 1.$$

91. Consider the following sinusoidal T.F.

$$G(j\omega) = \frac{(1 - 2j\omega)(1 + 5j\omega)}{(1 + j2\omega)(1 - j5\omega)}$$

Find the magnitude and phase $G(j\omega)$

(a)

$$G(j\omega) = 1 \underline{\left| 2(\tan^{-1} 5\omega + \tan^{-1} 2\omega) \right.}$$

(b)

$$G(j\omega) = 1 \underline{\left| 2(\tan^{-1} 5\omega - \tan^{-1} 2\omega) \right.}$$

(c)

$$G(j\omega) = 1 \underline{\left| 2(\tan^{-1} 2\omega + \tan^{-1} 5\omega) \right.}$$

(d)

$$G(j\omega) = 1 \underline{\left| -2(\tan^{-1} 2\omega + \tan^{-1} 5\omega) \right.}$$

92. **For the system described in question 91, what is the maximum phase contribution of $G(j\omega)$?**

(a) $\phi_m = -50.78°$.
(b) $\phi_m = +50.78°$.
(c) $\phi_m = -25.39°$.
(d) $\phi_m = +25.39°$.

93. **Consider the following characteristic polynomial.**

$$F(\lambda) = (\lambda + 2)(\lambda + 2 + j3)(\lambda + 2 - j3)$$

Determine whether the system is asymptotically and BIBO stable.

(a) Asymptotically stable but BIBO unstable.
(b) Asymptotically unstable but BIBO stable.
(c) Asymptotically and BIBO stable.
(d) Asymptotically and BIBO unstable.

94. **Consider the system as in question 93 with**

$$F(\lambda) = (\lambda - 2)(\lambda + 2 + j3)(\lambda + 2 - j3)$$

Find whether the system is BIBO and asymptotically stable.

(a) The system is BIBO and asymptotically stable.
(b) The system is BIBO and asymptotically unstable.
(c) The system is BIBO stable but asymptotically unstable.
(d) The system is asymptotically stable but BIBO unstable.

95. Consider the system whose characteristic polynomial is

$$F(\lambda) = (\lambda + 1)(\lambda^2 + 9)^2$$

Find whether the system is BIBO and asymptotically stable.

(a) The system is both asymptotically and BIBO stable.
(b) The system is both asymptotically and BIBO unstable.
(c) The system is BIBO stable but asymptotically unstable.
(d) The system is asymptotically stable but BIBO unstable.

96. Two sub-system $G_1(s)$ and $G_2(s)$ are connected in series.

$$g_1(t) = \delta(t) - 8e^{-4t}u(t)$$
$$g_2(t) = e^{4t}u(t)$$

Find whether the combined system is BIBO and asymptotically stable.

(a) A symbiotically and BIBO stable.
(b) A symbiotically and BIBO unstable.
(c) A symbiotically stable and BIBO unstable.
(d) A symbiotically unstable and BIBO stable.

97. The characteristic polynomial of a certain control system is given by

$$F(s) = 2s^4 + s^3 + 2s^2 + s + 2$$

Test whether the system is stable.

(a) The system is stable.
(b) The system is unstable.
(c) The system is marginally stable.
(d) None of the above.

98. Consider the following characteristic polynomial of a certain control system

$$F(s) = s^4 + 2s^3 + 6s^2 + Ks + 10$$

Find the value of K so that the system is stable.

(a) $K = 6 \pm j2$.
(b) $K = 6$.
(c) $K = \pm j2$.
(d) The system is unstable for any real value of K.

99. Consider the following characteristic polynomial.

$$F(s) = s^4 + 3s^3 + 12s^2 + 16s + K$$

Find the real value of K so that the system is stable.

(a) $0 < K < 35.56$.
(b) $0 < K < 52.92$.
(c) $0 < K < 76.59$.
(d) $0 < K < 88.76$.

100. The open loop T.F. of a certain unity feedback system is given by

$$G(s)\frac{K(s+10)}{s^3}$$

Find the real value of K so that the closed loop system is stable.

(a) $0 < K < 11.25$.
(b) $0 < K < 22.5$.
(c) $0 < K < 30.11$.
(d) The system is unstable for any real value of K.

Answers

(1) d
(2) c
(3) d
(4) d
(5) c
(6) c
(7) c
(8) c
(9) d
(10) d
(11) d
(12) d
(13) d
(14) d
(15) a
(16) b
(17) d
(18) d
(19) d
(20) d
(21) c
(22) c
(23) a
(24) a
(25) d
(26) a
(27) c
(28) b
(29) a
(30) b
(31) c
(32) c
(33) d
(34) d
(35) d
(36) b
(37) b
(38) b
(39) a
(40) b
(41) b
(42) b
(43) c
(44) d
(45) a
(46) b
(47) a
(48) b
(49) a
(50) a
(51) b
(52) a
(53) a
(54) c
(55) c
(56) c
(57) a
(58) c
(59) d
(60) a
(61) a
(62) c
(63) a
(64) b
(65) a
(66) c
(67) c
(68) b
(69) a
(70) b
(71) c
(72) c
(73) c
(74) c
(75) c
(76) a
(77) a
(78) c
(79) c
(80) c
(81) a
(82) c
(83) a
(84) c
(85) a
(86) a
(87) a
(88) a
(89) d
(90) c
(91) b
(92) b
(93) c
(94) b
(95) b
(96) d
(97) b
(98) d
(99) a
(100) d

Chapter 5
Root Locus Method of Analysis

Multiple Choice Questions

Choose the Correct Answer

1. **The root locus is the trajectory of**

   (a) the closed loop poles as the gain of the system K is varied in the range $0 \leq K \leq \infty$.
   (b) the closed loop poles as the gain K of the system K is varied in the range $-\infty < K < \infty$.
   (c) the closed loop zeros as the gain K is varied in the range $-\infty < K < \infty$.
   (d) poles and zeros as the gain K of the system is varied in the range $-\infty < K < \infty$.

2. **The criteria used to construct root locus are**

   (a) magnitude and phase angle criteria.
   (b) magnitude criterion.
   (c) phase angle criterion.
   (d) none of the above.

3. **How phase angle criterion is used to construct root locus for a negative feedback system for the +ve values of K?**

   (a) For any point s_1 in the complex s-plane to be a point in the root locus, the total sum of angles contributed by all the zeros should be zero.
   (b) The total sum of the angles drawn from all poles to s_1 point should be multiplies of 180°.
   (c) The difference of the angles of all the zeros and poles to the point s_1 should be odd multiple of 180°.

© The Author(s), under exclusive license to Springer Nature Switzerland AG 2026
S. Palani, *Multiple-Choice Questions in Control Systems Engineering*,
https://doi.org/10.1007/978-3-031-93787-3_5

(d) The angles of all the poles drawn to the point s_1 should be odd multiple of 180°.

4. How is magnitude criterion used in the root locus?

(a) The root locus is calibrated in terms of the product of all the distances drawn from poles of $GH(s)$.
(b) The root locus is calibrated in terms of the product of all the distances drawn from zeros of $GH(s)$.
(c) The root locus is calibrated in terms of K which is expressed as

$$K = \frac{\text{Product of directed distances drawn from all the poles of } GH(s) \text{ to the point } s_1}{\text{Product of directed distances drawn from all the zeros of } GH(s) \text{ to the point } s_1}.$$

(d) All the above.

5. The unique property of root locus is

(a) the root locus start from the poles of $GH(s)$ where $K = 0$ and end at the zeros or at infinity where K is ∞ at these points.
(b) the number of root locus for a given system is equal to the number of poles or zeros which ever is greater.
(c) magnitude criterion.
(d) phase angle criterion.

6. Root locus is a

(a) time domain approach.
(b) frequency domain approach.
(c) both time domain and frequency domain approach.
(d) none of them.

7. Root locus on the real axis is found

(a) if any point s_1 on the real axis has odd number of poles to the right of that point.
(b) if any point s_1 on the real axis has odd number of zeros to the right of that point.
(c) if any point s_1 on the real axis has odd number of poles and zeros to the right of that point.
(d) if any point s_1 has odd number of poles and zeros to the left of that point.

8. The centroid of the root locus is

(a) the intersect of the asymptotes on the real axis.
(b) center of gravity of the asymptotes.

(c) it gives guidance to the root locus.
(d) it is a measure of system stability.

9. The break-away and break-in points of root locus are

(a) when two and more than two branches of root-locus meet at one point and depart from that point, these points are called break-away or break-in points.
(b) if there is discontinuity in the root locus, these continuities are called break-away or break-in points.
(c) points at which $GH(s)$ has repeated poles are called break-away or break-in points.
(d) the points at which the T.F. $GH(s)$ has repeated zeros are called break-away or break-in points.

10. There are 5 poles 3 zeros in the T.F. $GH(s)$. The number of root loci are

(a) 8.
(b) 2.
(c) 5.
(d) 3.

11. The angle of the asymptote θ of the root locus is given by

(a)

$$\theta = \frac{(2n+1)\pi}{(p+z)}.$$

(b)

$$\theta = \frac{(2n+1)\pi}{(p-z)}.$$

(c)

$$\theta = \frac{\pi}{p+z}.$$

(d)

$$\theta = \frac{\pi}{(p-z)}.$$

12. Intersection of the asymptote (centroid) on the real axis is

(a)

$$\theta = \frac{[\sum \text{Rreal part of poles of } GH(s) - \sum \text{real part of zeros of } GH(s)]}{(p+z)}.$$

(b)

$$\theta = \frac{[\sum \text{real part of poles of } GH(s) - \sum \text{real part of zeros of } GH(s)]}{(p-z)}.$$

(c)

$$\theta = \frac{[\sum \text{real part of poles of } GH(s) + \sum \text{real part of zeros of } GH(s)]}{(p+z)}.$$

(d)

$$\theta = \frac{[\sum \text{real part of poles of } GH(s) + \sum \text{real part of zeros of } GH(s)]}{(p-z)}.$$

13. The break-away or break in points of the root locus are obtained by

(a) differentiating $GH(s)$ with respect to s and putting $dK/ds = 0$ and solving for s.
(b) differentiating $[1 - GH(s)]$ with respect to s and putting $dK/ds = 0$ and solving for s.
(c) differentiating $[1 + GH(s)]$ with respect to s and putting $dK/ds = 0$ and solving for s.
(d) differentiating denominator of $[1 + GH(s)]$ and putting $dK/ds = 0$ and solving for s.

14. Which one of the following statements is not correct?

(a) The root loci are continuous curves.
(b) Asymptotes some time fail to give proper guidance to the actual root locus.
(c) Root locus stat from zeros of $GH(s)(K = 0)$ and end at the poles of $GH(s)(K = \infty)$.
(d) Root locus does not exist for non-minimum phase T.F. of $GH(s)$.

15. Consider the following loop T.F. $GH(s)$

$$GH(s) = \frac{K}{s(s+1)(s+2)}.$$

The centroid σ and the critical gain K_c of the system are

(a) $\sigma = -1.5$; $K_c = 5$.
(b) $\sigma = -1.5$; $K_c = 6$.
(c) $\sigma = -2.0$; $K_c = 5$.
(d) $\sigma = -1$; $K_c = 6$.

16. Consider the following loop T.F. $GH(s)$ of a certain feedback system

$$GH(s) = \frac{K}{s(s+1)(s+2)(s+3)}.$$

The angles θ of the asymptotes σ and the break-away points are

(a) $\theta = \frac{\pi}{2}, \pi, \frac{3\pi}{2}$ and 2π; $\sigma = -0.2$ and -2.5.
(b) $\theta = \frac{\pi}{4}, \frac{3\pi}{2}, \frac{5\pi}{2}, \frac{7\pi}{2}$; $\sigma = -0.3$ and -2.5.
(c) $\theta = \frac{\pi}{6}, \frac{4\pi}{2}, \frac{7\pi}{2}, \frac{10\pi}{2}$; $\sigma = -0.3$ and -7.5.
(d) $\theta = 0, \frac{\pi}{2}, \pi, \frac{3\pi}{2}$; $\sigma = -0.3$ and -2.5.

17. A certain feedback control system has the following loop T.F.

$$GH(s) = \frac{K}{s(s+4)(s^2+4s+20)}.$$

For this the critical gain and the corresponding frequency of oscillations are

(a) $K_c = 200$; $\omega = \sqrt{10}$ rad/sec.
(b) $K_c = 260$; $\omega = 3$ rad/sec.
(c) $K_c = 2.65$; $\omega = 3.18$ rad/sec.
(d) $K_c = 260$; $\omega = \sqrt{10}$ rad/sec.

18. The loop T.F. of a certain feedback system is given by

$$GH(s) = \frac{K(s^2+2s+10)}{s(s+2)(s+4)}.$$

The values of break away points and the angle of departures are

(a) $s_b = 0.8$; $\theta_d = \pm 30°$.
(b) $s_b = 0.86$; $\theta_d = \pm 34°$.
(c) $s_b = 0.65$; $\theta_d = \pm 28°$.
(d) $s_b = 1.2$; $\theta_d = \pm 15°$.

19. In root locus the angle of arrival is calculated using

(a) $\theta_a = (2n+1)\pi - (\theta_p - \theta_z)$.
(b) $\theta_a = (2n-1)\pi - (\theta_p - \theta_z)$.
(c) $\theta_a = (2n+1)\pi + (\theta_p - \theta_z)$.
(d) $\theta_a = (2n-1)\pi - (\theta_p - \theta_z)$.

20. The angle of departure in root locus is calculated from

(a) $\theta_d = (2n + 1)\pi - (\theta_p - \theta_z)$.
(b) $\theta_d = (2n - 1)\pi - (\theta_p - \theta_z)$.
(c) $\theta_d = (2n + 1)\pi + (\theta_p - \theta_z)$.
(d) $\theta_d = (2n - 1)\pi + (\theta_p - \theta_z)$.

21. The unique purpose of drawing root locus is

(a) to determine the roots of the characteristic polynomial as the gain K is varied in the range $-\infty < K < \infty$.
(b) to find the open loop zeros and poles.
(c) to find the closed loop poles.
(d) to determine the stability of the system for any given value of K which is in the range $-\infty < K < \infty$.

22. How phase margin is determined from root locus plot? Find ω_1 in the $j\omega$ axis which satisfies the desired gain K_d. Let ϕ be the phase angle difference of $(\theta_z - \theta_p)$. The phase margin is given by

(a) $\phi_m = (180° - \phi)$.
(b) $\phi_m = (180° + \phi)$.
(c) $\phi_m = (90° - \phi)$.
(d) $\phi_m = (90° + \phi)$.

23. How gain margin is determined from root locus plot? If K_d is the desired gain and K_c is the critical gain, then gain margins is given by

(a) $G_m = \frac{K_d}{K_c}$.
(b) $G_m = \frac{K_c}{K_d}$.
(c) $G_m = K_c - K_d$.
(d) $G_m = K_d - K_c$.

24. Sketch the root locus for the following loop T.F. for $0 \leq K \leq \infty$ (Fig. 5.1).

$$GH(s) = \frac{K(s-2)}{(s+2)(s+3)}.$$

25. For the following loop T.F. draw the root locus (Fig. 5.2).

$$GH(s) = \frac{K(s+2)(s+4)}{(s-2)(s-4)}.$$

26. Consider the following open loop T.F. of a certain feedback system (Fig. 5.3).

$$GH(s) = \frac{K}{(s^2+4)}.$$

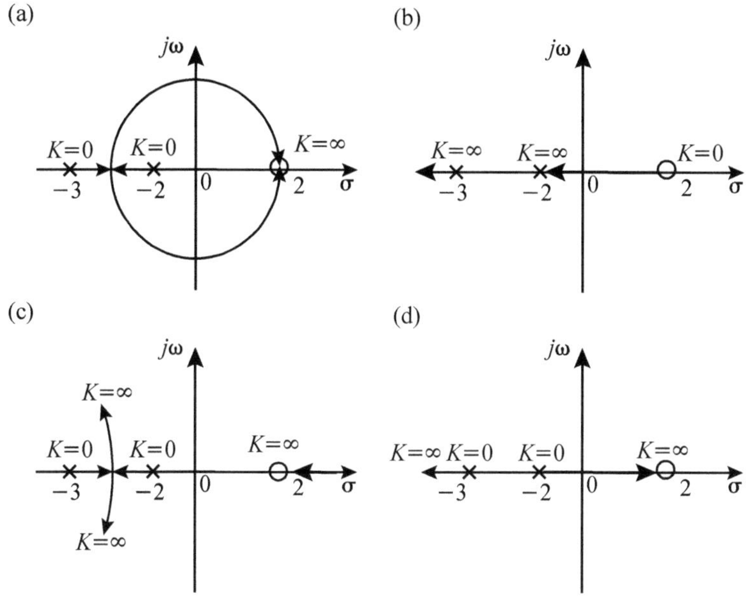

Fig. 5.1 Root locus for Question 24

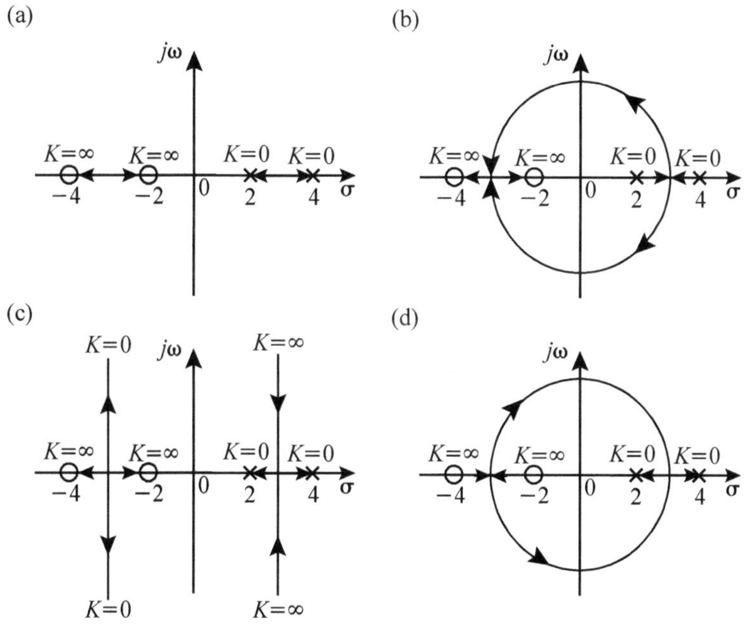

Fig. 5.2 Root locus for Question 25

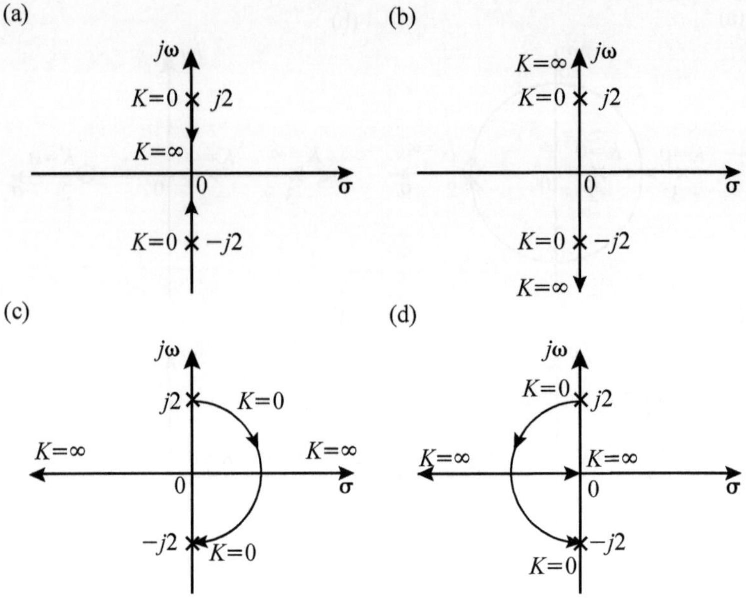

Fig. 5.3 Root locus for Question 26

27. Consider the following open loop T.F. of a certain feedback system (Fig. 5.4).

$$GH(s) = \frac{K(s+2)(s+4)}{(s^2+4)}.$$

Sketch the root locus for $0 \le K \le \infty$.

28. Consider the following open loop T.F. of a certain closed loop system (Fig. 5.5).

$$GH(s) = \frac{K}{s(s^2+4)}.$$

Sketch the root locus for $0 < K < \infty$.

29. Consider the following open loop T.F. of a certain closed loop system (Fig. 5.6).

$$GH(s) = \frac{K}{(s^2-4)}.$$

Sketch the root locus for $0 \le K \le \infty$.

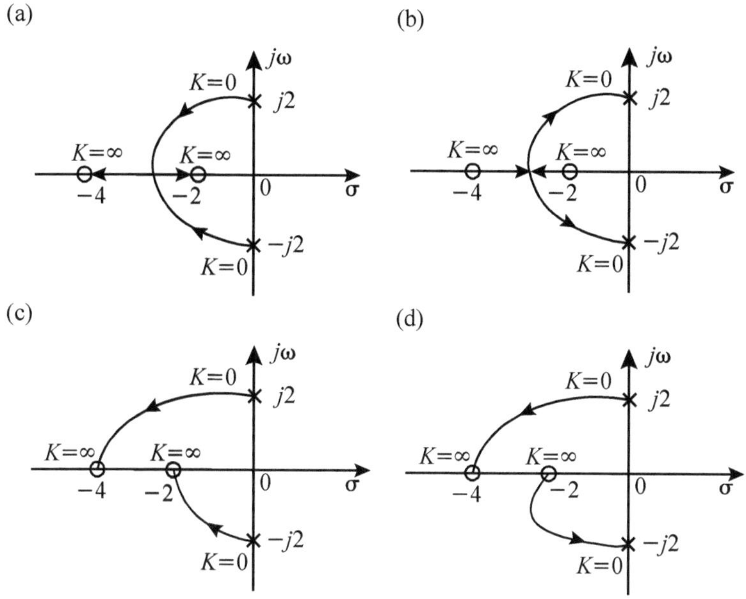

Fig. 5.4 Root locus for Question 27

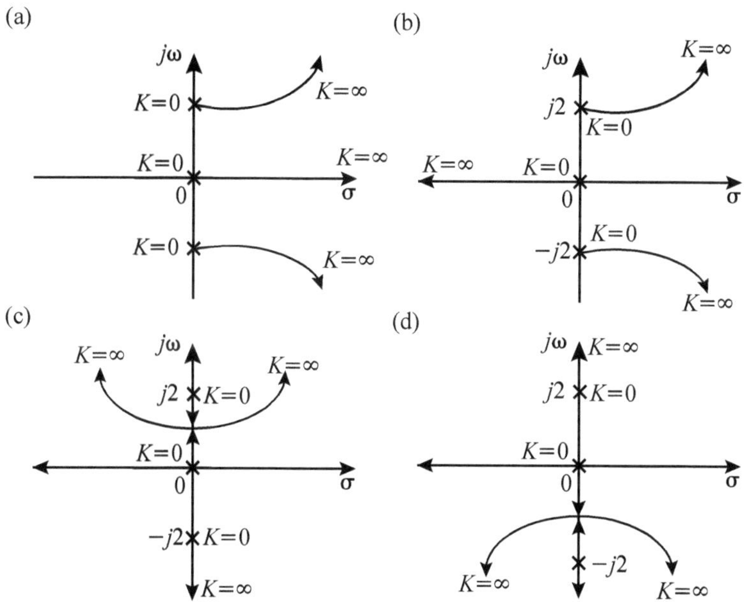

Fig. 5.5 Root locus for Question 28

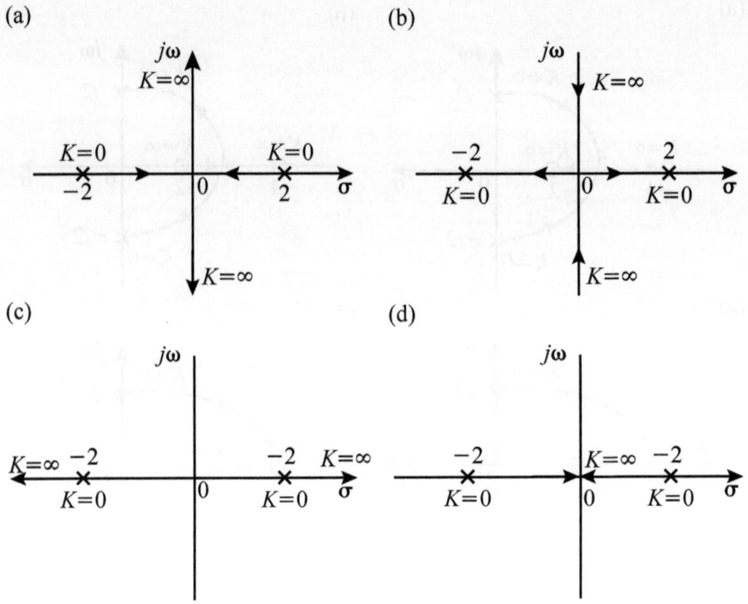

Fig. 5.6 Root locus for Question 29

30. A certain feedback control system has the following open loop T.F.

$$GH(s) = \frac{K(s+5)}{s(s+3)}.$$

Part of the root locus is found to be a circle. Find the centre and radius of this circle.

(a) Centre $(-5, 0)$ and radius $\sqrt{10}$.
(b) Centre $(-4, 0)$ and radius $\sqrt{10}$.
(c) Centre $(5, 0)$ and radius $\sqrt{10}$.
(d) Centre $(0, 0)$ and radius $\sqrt{10}$.

31. A certain feedback control system has the following open loop T.F.

$$GH(s) = \frac{K(s+A)}{s(s+B)}$$

Under what condition, part of the root locus is a circle for $0 \leq K \leq \infty$?

(a) $B > A$.
(b) $A > B$.
(c) $A = B$.
(d) All the above are possible.

32. In question 31, if $A < B$, what will be the nature of the root locus and what is the nature of damping?

(a) Root locus exists only on the negative real axis and the system is over damped.
(b) Root locus exists in the RHP only and the closed loop system is unstable.
(c) Part of root locus is a circle with centre $(A, 0)$ and radius $\sqrt{B}$.
(d) Since the part of root locus is circular and exists in LHP, the system is underdamped.

33. A certain feedback control system has the following open loop T.F.

$$GH(s) = \frac{K(s^2 + 2s + 10)}{s(s+2)(s+4)}.$$

The break-away point is

(a) -1.86.
(b) -1.2.
(c) -0.86.
(d) -0.5.

34. Consider the following open loop T.F.

$$G(s) = \frac{K(s+2)}{s(s+1)(s+4)}.$$

For what value of K, the system is critically damped?

(a) $K = 0.6$.
(b) $K = 0.4$.
(c) $K = 0.9$.
(d) $K = 0.1$.

35. The forward path T.F. of a certain control system is given by

$$GH(s) = \frac{K}{(s+4)(s^2 + 2s + 2)}.$$

Find the angle of departure of the root locus from the complex conjugate poles.

(a) $\theta_d = \pm 60°$.
(b) $\theta_d = \pm 67.67°$.
(c) $\theta_d = \pm 71.6°$.
(d) $\theta_d = \pm 75.22°$.

36. Consider the following open loop T.F. of a closed loop system

$$G(s) = \frac{K(s+2)}{s(s+1)(s+4)}.$$

The asymptotes of the root loci are

(a) $\pm 45°$ and $180°$ at $\sigma = -1.5$.
(b) $\pm 90°$ and $180°$ at $\sigma = -1.5$.
(c) $\pm 75°$ and $180°$ at $\sigma = -1.8$.
(d) $\pm 75°$ and $180°$ at $\sigma = -1.8$.

37. The characteristic polynomial of a certain system is given by

$$F(s) = s(s+4)(s^2+6s+10) + K.$$

The asymptotes and the critical value of the gain are

(a) $\pm 45°, \pm 135°; \sigma = -2.5; K_c = 120$.
(b) $\pm 45°, \pm 135°; \sigma = -2.5; K_c = 100$.
(c) $\pm 45°, \pm 135°; \sigma = -3; K_c = 100$.
(d) $\pm 45°, \pm 135°; \sigma = -3.5; K_c = 120$.

38. Consider the following open loop T.F. of a certain closed loop syste

$$G(s) = \frac{K(s+3)}{s(s+2)(s^2+4s+13)}.$$

The angles of departures of root loci from the complex poles are

(a) $\theta_d = \pm 79°$ at $s_{1,2} = -2 \pm j3$.
(b) $\theta_d = \pm 68°$ at $s_{1,2} = -2 \pm j3$.
(c) $\theta_d = \pm 61°$ at $s_{1,2} = -2 \pm j4$.
(d) $\theta_d = \pm 79°$ at $s_{1,2} = -2 \pm j4$.

39. Consider the system described in question 37. The break-away and break-in points are

(a) break-away point $= -1.2$; break-in point is 4.
(b) break-away point $= -4$; break-in point is -1.2.
(c) break-away point $= -1.2$; break-in point is -3.8.
(d) break-away point $= 3.8$; break-in point is -1.2.

40. Consider the open loop T.F. of a certain closed loop system

$$GH(s) = \frac{K(s^2+6s+25)}{s(s+3)(s^2+8s+25)}.$$

The break-away and break-in point of the root locus are

(a) break-away point $s_{1,2} = -2.8 \pm j5.4$ and break-in point is $s_3 = +1.3$.
(b) break-away point $s_{1,2} = -2.8 \pm j6.4$ and break-in point is $s_3 = -1.5$.
(c) break-away point $s_{1,2} = -5.4 \pm j2.8$ and break-in point is $s_3 = -1.3$.
(d) break-away point $s_{1,2} = -1.2, -1.5$ and break-in point is $s_3 = +1.3$.

41. What is angle criterion for root locus for $-\infty < K \leq 0$?

(a) $\phi|_{s1} =$ odd multiple of 180°.
(b) $\phi|_{s1} =$ even multiple of 180°.
(c) $\phi|_{s1} = s + z|_{s1}$ is 0°.
(d) $\phi\Big|_{s1} = \dfrac{1}{(s+p)}\Big|_{s1}$ is odd multiple of 180°.

42. What is angle criterion of a +ve feedback system?

(a) $\phi|_{s1} = (s + z)|_{s1}$ is 0°.
(b) $\phi\Big|_{s1} = \dfrac{1}{(s+p)}\Big|_{s1}$ is 180°.
(c) $\phi|_{s1} =$ even multiple of 180°.
(d) $\phi|_{s1} =$ odd multiple of 180°.

43. Consider the following open loop T.F. of a certain closed loop system

$$GH(s) = \frac{K(s+2)^2}{s^3}.$$

The critical value of gain K is

(a) $K_c > 1$.
(b) $K_c < 1$.
(c) $K_c < 10$.
(d) $K_c > 10$.

44. The open loop T.F. of a certain closed loop T.F. is given by

$$GH(s) = \frac{K(s+2)(s+6)}{s^2(s+4)}.$$

Find the value of K where $0 < K < \infty$ for which the break-in point of the locus is -10.3

(a) $K = 16.5$.
(b) $K = 18.7$.
(c) $K = 19.6$.
(d) $K = 21.2$.

45. The open loop T.F. of a certain closed loop T.F. is given by

$$GH(s) = \frac{K(s+2)}{(s^2+2s+2)}.$$

The value of K is in the range $0 < K < \infty$ the value of K for which the break-in point is -3.414 is

(a) $K = 10.28$.
(b) $K = 8.1$.
(c) $K = 4.828$.
(d) $K = 3.79$.

46. Consider the following T.F. of an open loop system

$$GH(s) = \frac{K}{s(s+1)(s+2)(s+3)}.$$

The toot locus of the closed loop system will have the centroid at,

(a) $\sigma = -0.25$.
(b) $\sigma = -1.5$.
(c) $\sigma = -0.2$.
(d) $\sigma = 1.5$.

47. The open loop T.F. of a certain closed loop T.F. is given by

$$GH(s) = \frac{K(1+2s)}{s^2(s+4.5)}.$$

The centroid and its angle are

(a) $\sigma = -2 \underline{|\pm 90^\circ}$.
(b) $\sigma = -2.5 \underline{|\pm 90^\circ}$.
(c) $\sigma = 2 \underline{|\pm 45^\circ}$.
(d) $\sigma = 2.5 \underline{|\pm 45^\circ}$.

48. The open loop T.F. of a certain closed loop system is given by

$$GH(s) = \frac{K(s+9)}{s(s^2+6s+10)}.$$

The centroid and its angle are

(a) $\sigma = -2.5 \underline{|\pm 90^\circ}$.
(b) $\sigma = +2.5 \underline{|\pm 90^\circ}$.
(c) $\sigma = -6.5 \underline{|\pm 90^\circ}$.
(d) $\sigma = 6.5 \underline{|\pm 90^\circ}$.

49. The open loop T.F. of a certain feedback control system is given by

$$GH(s) = \frac{K}{s(s+2)(s+5)}.$$

For the desired gain $K = 10$, what is the gain margin?

(a) $G_m = \frac{1}{\sqrt{10}}$.
(b) $G_m = 7$.
(c) $G_m = \sqrt{10}$.
(d) $G_m = \frac{1}{7}$.

50. Certain feedback control system has the following open loop T.F.

$$GH(s) = \frac{K(s+2)}{(s+1)^2}.$$

A part of root locus is a circle. The centre and radius of this circle are

(a) centre $c = (-1.5, 0)$ and $r = 1$.
(b) centre $c = (-1.8, 0)$ and $r = 1.5$.
(c) centre $c = (-2, 0)$ and $r = 1$.
(d) centre $c = (-2, 0)$ and $r = 2$.

51. Consider the following open loop T.F. of a closed loop system

$$GH(s) = \frac{K(s+2)(s^2+4)}{(s^2-16)}.$$

The value of K at the break point is

(a) $K = 0.126$.
(b) $K = 1.23$.
(c) $K = 5.72$.
(d) $K = 19.27$.

52. Consider the open loop T.F. of a certain closed loop system

$$GH(s) = \frac{K(s-2)(s-3)}{s(s+2)}.$$

Root locus plot shows that the critical value of gain $K_c = 4$. The closed loop poles are

(a) $s_{12} = \pm j$.
(b) $s_{12} = \pm j1.3$.
(c) $s_{12} = \pm j1.5$.
(d) $s_{12} = \pm j2$.

53. Consider the following T.F. of the open loop system

$$GH(s) = \frac{K(s-2)(s-3)}{(s+2)(s+3)}.$$

Find the centre and radius of the root locus part of which is circular.

(a) centre $c = (0, 0)$; $r = 2$.
(b) centre $c = (0, 0)$; $r = 2.5$.
(c) centre $c = (0, 0)$; $r = 3$.
(d) centre $c = (0, 0)$; $r = 4$.

54. Consider the following open loop T.F. of the certain closed loop system

$$GH(s) = \frac{Ks(s^2+9)}{(s+2)(s+3)(s^2+2s+2)}.$$

The critical value of K is

(a) $K = 0$.
(b) $K = \infty$.
(c) $K = 5$.
(d) $K = 10$.

55. The open loop T.F. of a certain feedback system is given by

$$GH(s) = \frac{K(s+5)}{s(s+3)}.$$

It is found that part of the root locus is a circle for $0 \leq K \leq \infty$. The maximum damping provided is

(a) $\varepsilon = 0.922$.
(b) $\varepsilon = 0.788$.
(c) $\varepsilon = 0.662$.
(d) $\varepsilon = 0.518$.

56. For the system described in question 55 the closed loop T.F. for $K = 0.675$, where the break-away point occur at $s = -1.84$.

(a)

$$T(s) = \frac{0.675(s+5)}{(s+1.84)^2}.$$

(b)

$$T(s) = \frac{(s+5)}{(s+1.84)^2}.$$

(c)

$$T(s) = \frac{1}{(s + 1.84)^2}.$$

(d)

$$T(s) = \frac{1}{(s + 1.84)^2}.$$

57. The open loop T.F. of a certain unity negative feedback system is given by

$$GH(s) = \frac{4}{(s^3 - 64)}.$$

The break-away points are at

(a) $s = (-1, 0)$.
(b) $s = (+1, 0)$.
(c) $s = (0, 0)$.
(d) $s = (-1.5, 0)$.

58. For the question 57, the system gain at the break away point is

(a) $K = 52$.
(b) $K = 59$.
(c) $K = 64$.
(d) $K = 70$.

59. For the question 57, how many closed loop roots are there in LHP and RHP for $K < 64$?

(a) LHP roots 1 and RHP roots 2.
(b) LHP roots 2 and RHP roots 1.
(c) LHP roots 3 and RHP roots 0.
(d) LHP roots 0 and RHP roots 3.

60. For the question 57, how many closed loop roots are there in LHP and RHP for $K > 64$?

(a) LHP roots 1 and RHP roots 2.
(b) LHP roots 2 and RHP roots 1.
(c) LHP roots 3 and RHP roots 0.
(d) LHP roots 0 and RHP roots 3.

61. For the question 57, for $K = 64$, what is the closed loop T.F.

(a)

$$\frac{C(s)}{R(s)} = \frac{64}{s^3 + 64}.$$

(b)

$$\frac{C(s)}{R(s)} = \frac{64}{s^3 - 64}.$$

(c)

$$\frac{C(s)}{R(s)} = \frac{64}{s^3}.$$

(d) None of the above.

62. Consider the following open loop T.F. of a certain unity feedback system.

$$G(s) = \frac{K(s+2)}{(s^2+2s+2)}.$$

The break in point of the root locus is at $s = -3.414$ and the corresponding value of $K = 4.82$. For this value of K, the closed loop T.F. is

(a)

$$\frac{C(s)}{R(s)} = \frac{4.82}{(s+3.414)}.$$

(b)

$$\frac{C(s)}{R(s)} = \frac{4.82}{(s+3.414)^2}.$$

(c)

$$\frac{C(s)}{R(s)} = \frac{4.82(s+2)}{(s+3.414)}.$$

(d)

$$\frac{C(s)}{R(s)} = \frac{4.82(s+2)}{(s+3.414)^2}.$$

63. A certain +ve feedback control system has the following open loop T.F.

$$G(s) = \frac{K(s+9)}{s(s^2+4s+1)}.$$

From the root locus plot it is observed that the break in point is at $s = -13.3$ and the corresponding K value is $K = 4.20$. At $K = 420$ there is a closed loop pole at $s = 22$ in the root locus. Find the closed loop T.F.

(a)

$$\frac{C(s)}{R(s)} = \frac{420}{(s+13.3)(s-22)}.$$

(b)

$$\frac{C(s)}{R(s)} = \frac{420}{(s + 13.3)^2(s - 22)}.$$

(c)

$$\frac{C(s)}{R(s)} = \frac{420(s + 9)}{(s + 13.3)^2(s - 22)}.$$

(d)

$$\frac{C(s)}{R(s)} = \frac{420(s + 9)}{(s + 13.3)(s - 22)}.$$

64. A certain unity negative feedback system has the following open loop T.F.

$$G(s) = \frac{K(s + 2)}{(s + 1)^2}.$$

From the root locus plot it is observed that part of root locus is a circle. The break in point observed is at $s = -3$ where the value of $K = 4$. Find the centre and radius of the root locus.

(a) Centre $c = (-1, 0)$; $r = 1$.
(b) Centre $c = (-2, 0)$; $r = 1$.
(c) Centre $c = (-2, 0)$; $r = 2$.
(d) Centre $c = (+2, 0)$; $r = 1$.

65. For question 64, what is the closed loop T.F. for $K = 4$?

(a)

$$\frac{C(s)}{R(s)} = \frac{4(s + 2)}{(s + 3)}.$$

(b)

$$\frac{C(s)}{R(s)} = \frac{4}{(s + 3)}.$$

(c)

$$\frac{C(s)}{R(s)} = \frac{4}{(s + 3)^2}.$$

(d)

$$\frac{C(s)}{R(s)} = \frac{4(s + 2)}{(s + 3)^2}.$$

66. A certain negative unity feedback control system has the following open loop T.F

$$GH(s) = \frac{K}{(s^4 - 81)}.$$

Determine the locations of the open loop system poles in the s-plane

(a) $(s + 1)(s - 1)(s + j1)(s + j1)$.
(b) $(s + 3)(s - 3)(s + j1)(s - j1)$.
(c) $(s + 3)(s - 3)(s + j3)(s - j3)$.
(d) $(s + 1)(s + 1)(s + j3)(s - j3)$.

67. For the question 66, find break in points of the root locus.

(a) $s = 0$.
(b) $s = 3$.
(c) $s = -3$.
(d) None of the above.

68. For the question 66, find the value of K at the break in point.

(a) $K = 3$.
(b) $K = 9$.
(c) $K = 27$.
(d) $K = 81$.

69. For question 66, what is the closed loop T.F. for $K = 81$?

(a)

$$\frac{C(s)}{R(s)} = \frac{81}{(s^4 - 81)}.$$

(b)

$$\frac{C(s)}{R(s)} = \frac{81}{s^4}.$$

(c)

$$\frac{C(s)}{R(s)} = \frac{1}{s^4}.$$

(d)

$$\frac{C(s)}{R(s)} = \frac{1}{(s^4 - 81)}.$$

70. The forward path T.F. of a certain unity feedback control system is given by

$$G(s) = -\frac{K}{(s^3 + 64)}.$$

Find the location of the poles of the open loop T.F.

(a) $s_1 = -3;\ s_{2,3} = -2 \pm j3.46.$
(b) $s_1 = -3;\ s_{2,3} = -2 \pm j3.46.$
(c) $s_1 = -3;\ s_{2,3} = 2 \pm j3.46.$
(d) $s_1 = 3;\ s_{2,3} = 2 \pm j3.46.$

71. **A certain unity feedback control system has the following open loop T.F.**

$$GH(s) = -\frac{K(s+2)(s^2+4)}{(s^2-16)}.$$

Locate the open loop poles and zeros in the s-plane.

(a) $z_1 = -2;\ z_{2,3} = \pm j\sqrt{2};\ p_1 = 4;\ p_2 = 4.$
(b) $z_1 = -2;\ z_{2,3} = \pm j\sqrt{2};\ p_1 = 4;\ p_2 = -4.$
(c) $z_1 = 2;\ z_{2,3} = \pm j\sqrt{2};\ p_1 = 4;\ p_2 = 4.$
(d) $z_1 = -2;\ z_{2,3} = \pm j\sqrt{2};\ p_1 = -4;\ p_2 = -4.$

72. **For the system in question 71, find the breakaway point.**

(a) −5.27.
(b) −6.67.
(c) −7.23.
(d) −8.15.

73. **Consider the following open loop T.F. of a certain unity feedback system**

$$G(s) = \frac{K(s-2)(s-3)}{s(s+2)}.$$

It is found that part of the root locus is a circle. The centre and radius of this circle are

(a) $c = 0.86;\ r = 1.15.$
(b) $c = 1.15;\ r = 0.86.$
(c) $c = 1.55;\ r = 0.86.$
(d) $c = 0.86;\ r = 1.55.$

74. **For the question 73 find the breakaway point.**

(a) −0.51.
(b) +0.51.
(c) −0.69.
(d) +0.69.

75. **For the question 74 find the break in point.**

(a) +2.41.
(b) −2.41.

(c) −0.69.
(d) +0.69.

76. For the question 73, find the value of K for zero damping. What is the frequency of oscillation?

(a) $K_c = 0.4$; $\omega = 1.3$ rad/sec.
(b) $K_c = 1.3$; $\omega = 0.4$ rad/sec.
(c) $K_c = 0.5$; $\omega = 1.3$ rad/sec.
(d) $K_c = 0.4$; $\omega = 1.3$ rad/sec.

77. For the system in question 73, the breakaway point is at −0.69 where the value of $K = 0.091$. Find the closed loop T.F.

(a)
$$\frac{C(s)}{R(s)} = \frac{0.091}{(s+0.69)^2}.$$

(b)
$$\frac{C(s)}{R(s)} = \frac{0.91(s-2)(s-3)}{(s+0.69)^2}.$$

(c)
$$\frac{C(s)}{R(s)} = \frac{0.0834(s-2)(s-3)}{(s+0.69)^2}.$$

(d)
$$\frac{C(s)}{R(s)} = \frac{0.0834(s-2)(s+3)}{(s+0.69)^2}.$$

78. For the system in question 73, the break in point is $s = 2.41$ and the corresponding $K = 44$. Find the closed loop T.F.

(a)
$$\frac{C(s)}{R(s)} = \frac{44(s-2)(s-3)}{(s+2.41)^2}.$$

(b)
$$\frac{C(s)}{R(s)} = \frac{0.978(s-2)(s-3)}{(s+2.41)^2}.$$

(c)
$$\frac{C(s)}{R(s)} = \frac{0.988(s-2)(s-3)}{(s+2.41)}.$$

(d)
$$\frac{C(s)}{R(s)} = \frac{44(s-2)(s-3)}{(s+2.41)^2}.$$

79. A certain unity negative feedback system has the following open loop T.F.

$$G = \frac{K(s^2 - 4s + 5)}{(s+1)(s+2)}.$$

It is found that part of the root locus is a circle. Find the centre and radius of the circle.

(a) $c = -0.38;\ r = 1.82.$
(b) $c = +0.38;\ r = 1.82.$
(c) $c = 0;\ r = 2.$
(d) $c = 1.82;\ r = 0.38.$

80. For the system described in question 79, find the value of K for zero damping and the corresponding frequency.

(a) $K_c = 0.75;\ \omega = 1.81$ rad/sec.
(b) $K_c = 0.62;\ \omega = 1.7$ rad/sec.
(c) $K_c = 1.0;\ \omega = 2.0$ rad/sec.
(d) $K_c = 2.5;\ \omega = 5$ rad/sec.

81. For the system described in question 79, find the breakaway point.

(a) $s = -1.88.$
(b) $s = -1.44.$
(c) $s = +1.88.$
(d) $s = +1.44.$

82. For the system described in question 79, the breakaway point is at $s = -1.44$ and the corresponding gain $K = 0.0192$. The closed loop T.F. is

(a)

$$\frac{C(s)}{R(s)} = \frac{0.0192(s^2 - 4s + 5)}{(s + 1.44)^2}.$$

(b)

$$\frac{C(s)}{R(s)} = \frac{(s^2 + 4s + 5)}{(s + 1.44)^2}.$$

(c)

$$\frac{C(s)}{R(s)} = \frac{0.0188(s^2 - 4s + 5)}{(s + 1.44)^2}.$$

(d)

$$\frac{C(s)}{R(s)} = \frac{0.0188}{(s + 1.44)^2}.$$

83. A certain unity negative feedback system has the following open loop T.F.

$$G = \frac{K(s-2)(s-3)}{(s+2)(s+3)}.$$

It is found that part of the root locus is a circle. Find the centre and radius of the circular root locus.

(a) $c = -1.0;\ r = 2.5$.
(b) $c = +1.08;\ r = 2.5$.
(c) $c = 0;\ r = 3$.
(d) $c = 0;\ r = 2.5$.

84. For the system described in question 83, find the value of K for zero damping and the corresponding frequency of oscillation.

(a) $K_c = 2.15;\ \omega = 1.5$ rad/sec.
(b) $K_c = 1.0;\ \omega = 2.15$ rad/sec.
(c) $K_c = 1.0;\ \omega = 2.5$ rad/sec.
(d) $K_c = 2.15;\ \omega = 2.45$ rad/sec.

85. For the system described in question 83, find the breakaway point.

(a) $s = 2.5$.
(b) $s = -2.5$.
(c) $s = 3.0$.
(d) $s = -3.5$.

86. For the system described in question 83, the breakaway point is at $s = -2.5$. Find the corresponding value of K.

(a) $K = 100$.
(b) $K = 10$.
(c) $K = 0.1$.
(d) $K = 0.01$.

87. For the system described in question 83, the breakaway point is at $s = -2.5$ and the corresponding value of $K = 0.01$. For this value of K, find the closed loop T.F.

(a)

$$\frac{C(s)}{R(s)} = \frac{0.01(s-2)(s-3)}{(s+2.5)^2}.$$

(b)

$$\frac{C(s)}{R(s)} = \frac{0.01(s-2)(s-3)}{(s+2.5)}.$$

(c)

$$\frac{C(s)}{R(s)} = \frac{0.0099(s-2)(s-3)}{(s+2.5)}.$$

(d)

$$\frac{C(s)}{R(s)} = \frac{0.0099(s-2)(s-3)}{(s+2.5)}.$$

88. For the system described in question 83, find the break in point.

(a) $s = -2.5$.
(b) $s = +2.5$.
(c) $s = -2.0$.
(d) $s = +2.0$.

89. For the system described in question 83, find the value of K at the break in point.

(a) $K = 0.1$.
(b) $K = 1.0$.
(c) $K = 10$.
(d) $K = 100$.

90. For the system described in question 83, the breakaway point is at $s = 2.5$ and the corresponding value of $K = 100$. Find the closed loop T.F.

(a)

$$\frac{C(s)}{R(s)} = \frac{100(s-2)(s-3)}{(s-2.5)^2}.$$

(b)

$$\frac{C(s)}{R(s)} = \frac{0.99(s-2)(s-3)}{(s-2.5)^2}.$$

(c)

$$\frac{C(s)}{R(s)} = \frac{0.99(s-2)(s-3)}{(s+2.5)^2}.$$

(d)

$$\frac{C(s)}{R(s)} = \frac{100(s-2)(s-3)}{(s+2.5)^2}.$$

91. Certain negative feedback system has the following open loop T.F.

$$G(s) = \frac{K(s+2)(s+3)}{(s-2)(s-3)}.$$

It is observed that part of root locus is a circle. Find the centre and radius of the circular root locus.

(a) $c = 0;\ r = 2.5$.
(b) $c = -1;\ r = 2.5$.
(c) $c = +1;\ r = 2.5$.
(d) $c = 0;\ r = 3.0$.

92. For the system described in question 91, find the break away point.

(a) $c = 2.5$.
(b) $c = -2.5$.
(c) $c = 2.0$.
(d) $c = -2.0$.

93. For the system described in question 91, the breakaway point is found at $s = +2.5$. Find the value of K at this point.

(a) $K = 100$.
(b) $K = 10$.
(c) $K = 0.1$.
(d) $K = 0.01$.

94. For the system described in question 91, the breakaway point is observed at $s = 2.5$ and the corresponding gain $K = 0.01$. Find the closed loop T.F.

(a)
$$\frac{C(s)}{R(s)} = \frac{0.01(s+2)(s+3)}{(s-2.5)^2}.$$

(b)
$$\frac{C(s)}{R(s)} = \frac{0.0099(s+2)(s+3)}{(s-2.5)^2}.$$

(c)
$$\frac{C(s)}{R(s)} = \frac{0.01(s+2)(s+3)}{(s+2.5)^2}.$$

(d)
$$\frac{C(s)}{R(s)} = \frac{0.01(s+2)(s+3)}{(s+2.5)^2}.$$

95. For the system described in question 91, find the break in point.

(a) $s = 2.5$.
(b) $s = -2.5$.
(c) $s = 3.0$.
(d) $s = -3.5$.

96. For the system described in question 91, find the value of the gain K at the break in point $s = -2.5$.

(a) $K = 0.1$.
(b) $K = 1.0$.
(c) $K = 10.0$.
(d) $K = 100.0$.

97. For the system described in question 91, it is observed that the break in point is at $s = -2.5$ and the corresponding gain $K = 100$. Find the closed loop T.F.

(a)

$$\frac{C(s)}{R(s)} = \frac{100(s+2)(s+3)}{(s+2.5)^2}.$$

(b)

$$\frac{C(s)}{R(s)} = \frac{0.99(s+2)(s+3)}{(s+2.5)^2}.$$

(c)

$$\frac{C(s)}{R(s)} = \frac{100(s+2)(s+3)}{(s-2.5)^2}.$$

(d)

$$\frac{C(s)}{R(s)} = \frac{0.99(s+2)(s+3)}{(s-2.5)^2}.$$

98. For the system described in question 91, find the value of K for which the damping is zero. Also find the corresponding frequency of oscillation.

(a) $K_c = 1;\ \omega = 2.5$ rad/sec.
(b) $K_c = 1.5;\ \omega = 2.5$ rad/sec.
(c) $K_c = 2.0;\ \omega = 3.0$ rad/sec.
(d) $K_c = 2.5;\ \omega = 3.5$ rad/sec.

99. A certain unity negative feedback system has the following open loop T.F.

$$G(s) = \frac{K(s^2-16)}{(s^2+9)}.$$

Find the break in point and the corresponding value of the gain K.

(a) $s = 0;\ K = -0.5$.
(b) $s = 0;\ K = -0.565$.
(c) $s = 0;\ K = -0.6$.
(d) $s = 0;\ K = -0.656$.

100. For the system described in question 99, the break in point is at $s = 0$ and the corresponding K value is 0.565. Find the closed loop T.F.

(a)
$$\frac{C(s)}{R(s)} = \frac{0.565(s^2 - 16)}{s^2}.$$

(b)
$$\frac{C(s)}{R(s)} = \frac{0.361(s^2 - 16)}{s^2}.$$

(c)
$$\frac{C(s)}{R(s)} = \frac{0.361(s^2 - 16)}{s^2 - 2}.$$

(d)
$$\frac{C(s)}{R(s)} = \frac{0.565(s - 16)}{(s - 2)^2}.$$

Answers

(1) b
(2) a
(3) c
(4) c
(5) d
(6) c
(7) c
(8) a
(9) a
(10) c
(11) b
(12) b
(13) c
(14) c
(15) d
(16) b
(17) d
(18) b
(19) c
(20) a
(21) a
(22) a
(23) b
(24) d
(25) b
(26) b
(27) a
(28) b
(29) a
(30) a
(31) b
(32) a
(33) c
(34) a
(35) c
(36) b
(37) a
(38) a
(39) c
(40) a
(41) b
(42) c
(43) a
(44) b
(45) c
(46) b
(47) a
(48) b
(49) a
(50) c
(51) a
(52) b
(53) b
(54) b
(55) b
(56) a
(57) c
(58) c
(59) b
(60) a
(61) c
(62) d
(63) c
(64) b
(65) d
(66) c
(67) a
(68) d
(69) b
(70) c
(71) b
(72) b
(73) d
(74) c
(75) a
(76) a

(77) c
(78) b
(79) b
(80) a
(81) b
(82) c
(83) d
(84) c
(85) b
(86) d
(87) d
(88) b
(89) d
(90) b
(91) a
(92) a
(93) d
(94) b
(95) b
(96) d
(97) b
(98) a
(99) b
(100) b

Chapter 6
Design of Compensators

Multiple Choice Questions

Choose the Correct Answer

1. **What are compensators?**
   (a) Even a well designed plant when put into piratical applications may not give prescribed standards. In such cases devices are introduced along with the plant or process. Such devices are called compensators.
   (b) Devices which meet the required standards are called compensators.
   (c) Devices along with the input which give satisfactory performance are called compensators.
   (d) All of the above.
2. **In what configurations the compensators are connected?**
   (a) Series compensators.
   (b) Parallel compensators.
   (c) Series-parallel compensators.
   (d) All the above.
3. **What is a controller?**
   (a) The input or the difference between the input and output is applied to a device which sends signal to the plant to maintain the desired standard. Such device is called controller.
   (b) The device which is connected in cascade of the plant is a controller.
   (c) The device which is connected in parallel with the plant is a controller.
   (d) Devices which are connected in series, parallel combination and connected with the plant are called a controllers.

© The Author(s), under exclusive license to Springer Nature Switzerland AG 2026
S. Palani, *Multiple-Choice Questions in Control Systems Engineering*,
https://doi.org/10.1007/978-3-031-93787-3_6

4. The devices used as controller are,

(a) lever arrangement.
(b) an amplifier.
(c) a computer.
(d) all the above.

5. The device used as compensator are

(a) lead compensator.
(b) lag compensator.
(c) lag-lead compensator.
(d) all the above.

6. Compensators and controllers are designed in

(a) time domain.
(b) frequency domain.
(c) time and frequency domain.
(d) any one of time domain or frequency domain.

7. What are the time domain specifications used to design controllers and compensators?

(a) Peak over-shoot.
(b) Rise time.
(c) Settling time.
(d) All the above three.

8. What are the frequency domain specifications used to design controllers and compensators?

(a) Resonance peak.
(b) Bandwidth.
(c) Gain margin and phase margin.
(d) All the above.

9. What methods are used to design controllers and compensators in the frequency domain?

(a) Polar plot.
(b) Bode plot.
(c) Log magnitude-phase plot.
(d) All the above one at a time.

10. What methods are used in the time domain to design controllers and compensators?

(a) Root locus method.
(b) Time response of the system.
(c) Polar plot.

(d) Log magnitude-phase plot.

11. Is it possible to determine the frequency domain specifications such as G_m, ϕ_m, ω_{gc} and ω_{pc} from root locus which is a time domain plot?

(a) G_m and ω_{pc} can be found out.
(b) ϕ_m and ω_{gc} can be found out.
(c) G_m, ϕ_m, ω_{gc} and ω_{pc} can be found out.
(d) None of these things can be found out from root locus.

12. What is the unique advantage of series compensator?

(a) The series compensator is simpler.
(b) It requires less equipment and hence cost is low.
(c) The energy flow is from lower level to higher level and hence no amplifier is necessary.
(d) All the above.

13. What is the unique disadvantage of series compensator?

(a) The energy flow is from lower level to higher level and an amplifier is required.
(b) It requires more components and hence cost is high.
(c) Designing a series compensator is a difficult process.
(d) All the above.

14. What is the unique advantage of parallel compensator?

(a) The energy flow is from higher level to lower level which requires no amplifier.
(b) The number of equipment required is less and cost is low.
(c) Design process is simpler.
(d) All the above.

15. The root locus method is preferred to design compensators because

(a) the poles and zeros can be easily adjusted as per requirements in the root locus method.
(b) using computational methods such as MATLAB and Pyson, the compensator parameters can be easily determined.
(c) the transient response specifications are easily identified using root locus.
(d) even the frequency response specifications can be identified in root locus.

16. The stability and transient response depend upon pole location only. Justify?

(a) The transient response depends upon mainly by pole locations but zero also affects the transient response.
(b) The stability is decided by the poles and zeros of the closed loop system.
(c) Stability is decided by the locations of the poles and zeros of the closed loop T.F.
(d) None of the three is correct.

17. The T.F. of a lead network is

(a)

$$G_c(s) = \frac{\alpha(1+Ts)}{(1+\alpha Ts)};\ \alpha > 1.$$

(b)

$$G_c(s) = \frac{\alpha(1+Ts)}{(1+\alpha Ts)};\ \alpha < 1.$$

(c)

$$G_c(s) = \frac{\alpha(1+\alpha Ts)}{(1+Ts)};\ \alpha < 1.$$

(d)

$$G_c(s) = \frac{\alpha(1+\alpha Ts)}{(1+Ts)};\ \alpha > 1.$$

18. The polar plot of a lead network is, (Fig. 6.1)

19. Draw the Bode plot of a lead network (Fig. 6.2).

20. The structure of the lead network and its pole-zero locations are (Fig. 6.3)

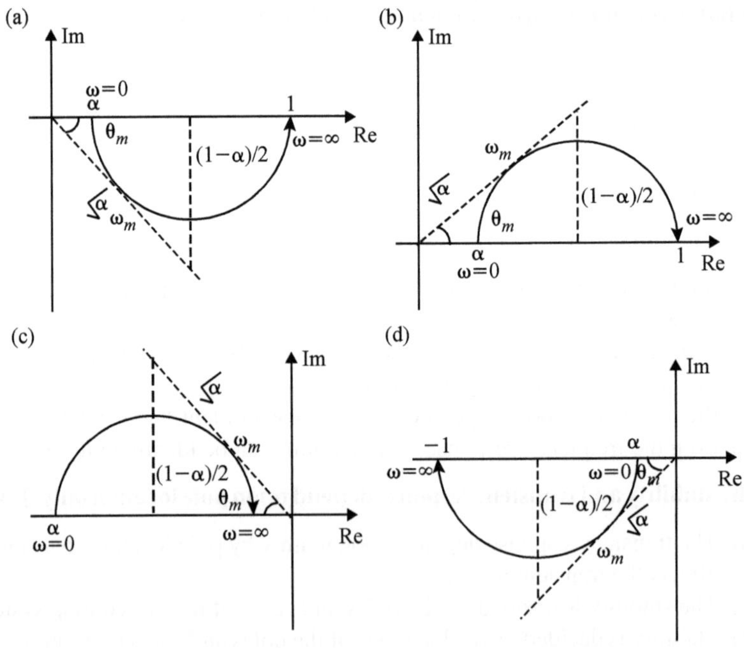

Fig. 6.1 Polar plot of a lead network

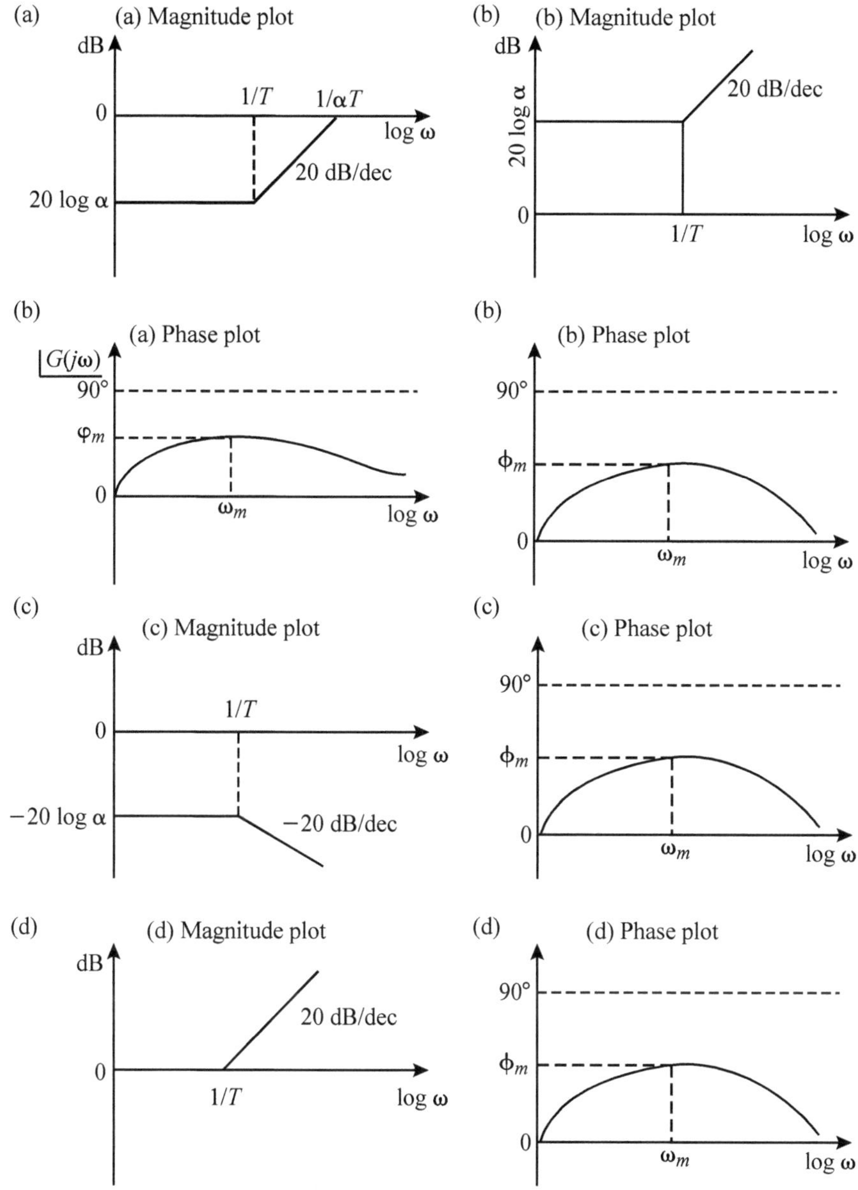

Fig. 6.2 Bode magnitude and phase plot of a lead compensator

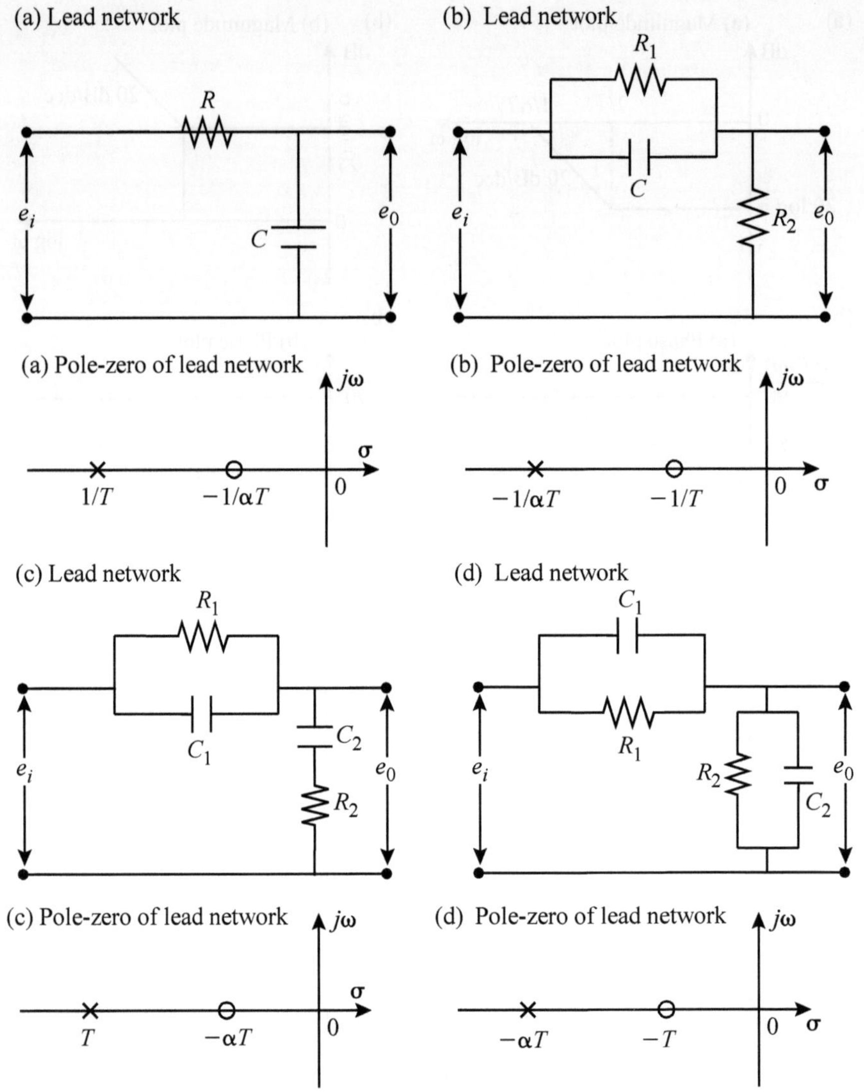

Fig. 6.3 Structure of electrical lead network and the pole zero diagrams

21. From a lead network the α is given by

(a)
$$\alpha = \frac{(1 + \sin\phi_m)}{(1 - \sin\phi_m)}.$$

(b)
$$\alpha = \frac{(\sin\phi_m - 1)}{(1 + \sin\phi_m)}.$$

(c)
$$\alpha = \frac{(1 - \sin\phi_m)}{(\sin\phi_m - 1)}.$$

(d)
$$\alpha = \frac{(1 - \sin\phi_m)}{(1 + \sin\phi_m)}.$$

22. From a lead network ω_m is given by

(a) $\omega_m = \frac{\sqrt{\alpha}}{T}$.
(b) $\omega_m = \frac{1}{T\sqrt{\alpha}}$.
(c) $\omega_m = T\sqrt{\alpha}$.
(d) $\omega_m = \frac{T}{\sqrt{\alpha}}$.

23. How does a phase lead network perform?

(a) The phase lead network offers excessive phase lead and offsets phase lag which is inherently present in the system.
(b) The lead network is designed such that at the gain cross-over frequency the maximum phase contribution comes from the lead network which increases the phase margin.
(c) As for as possible the altercation factors α should be low and practically realizable. This increases the gain margin. High value of $1/\alpha$ increases noise.
(d) All the above three methods.

24. What is the effect of steady state error when a lead network is employed?

(a) The lead network increases the steady state error.
(b) The lead network decreases the steady state error.
(c) Since lead network provides attenuation at low frequency, it will not have much impact in the steady state error.
(d) Depending upon the choice of the attenuation factor α, the steady state error will change.

25. What is the maximum phase lead that can be practically obtained from a lead compensator?

(a) $40 \leq \phi_l \leq 45°$.
(b) $45° \leq \phi_l \leq 50$.
(c) $50° \leq \phi_l \leq 60°$.
(d) $60° \leq \phi_l \leq 70°$.

26. What will be the effect of lead network on the system bandwidth, rise time and settling time?

(a) Bandwidth, rise time and settling time are all increased.
(b) Bandwidth is reduced and rise time and settling time increased.
(c) Bandwidth is increased and rise time and settling time are all reduced.
(d) Bandwidth, rise time and settling time are all reduced.

27. What will be limiting factor of the attenuation factor α when a lead network is introduced?

(a) $0.01 \leq \alpha < 0.02$.
(b) $0.02 \leq \alpha \leq 0.03$.
(c) $0.03 \leq \alpha \leq 0.05$.
(d) $\alpha \leq 0.07$.

28. A certain system has more than 60° phase leg. How to design a lead network?

(a) A lead network is designed with an amplifier.
(b) Two phase lead networks each one contributing a phase lead of more than 30° are connected in cascade.
(c) Two phase lead networks each one contributing a phase lead of more than 30° each are connected in cascade with an amplifier connected in between them.
(d) Lead network cannot be designed for this network.

29. The structure of a lag compensator is, (Fig. 6.4)
30. The T.F. of a lag network is

(a)

$$G_c(s) = \frac{(1+\alpha Ts)}{(1+Ts)}.$$

(b)

$$G_c(s) = \frac{(1+Ts)}{(1+\alpha Ts)}.$$

(c)

$$G_c(s) = \frac{\alpha(1+Ts)}{(1+\alpha Ts)}.$$

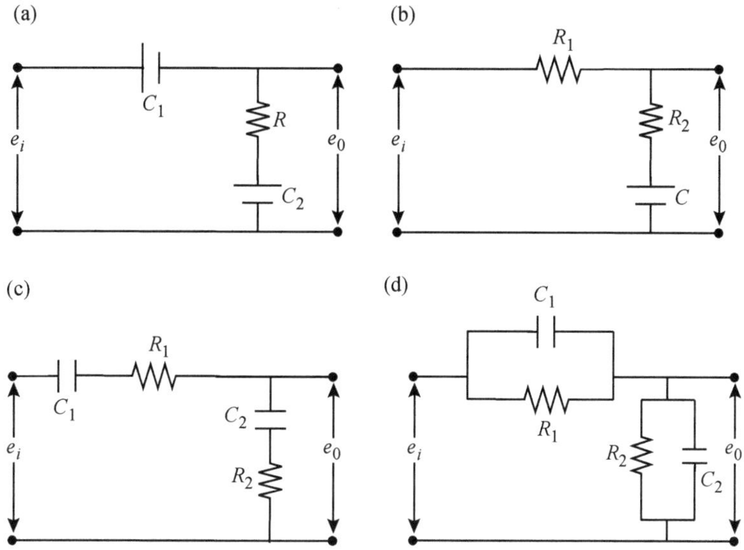

Fig. 6.4 Lag compensator network

(d)

$$G(s) = \frac{\alpha(1 + \alpha T s)}{(1 + T s)}.$$

31. The value of the attenuation factor is for a lag network is

(a) $\alpha > 1$.
(b) $\alpha < 1$.
(c) $\alpha < -1$.
(d) $\alpha > -1$.

32. Sketch the Bode plot for a leg compensator (Fig. 6.5).

33. The frequency regions of a compensators are divided into

(a) low frequency region.
(b) middle frequency region.
(c) high frequency region.
(d) all the three.

34. What does low frequency region represent?

(a) Low frequency region represents the steady state response of the system.
(b) Middle frequency region represents the transient state response.
(c) High frequency region represents stability of the system.
(d) None of the above.

35. What response of the Bode frequency plot represents the stability of the system?

(a) The low frequency response.
(b) Middle frequency response.
(c) High frequency response.
(d) Very high frequency response.

36. What frequency region represents the transient response of the system?

(a) Low frequency response region.
(b) Middle frequency response region.
(c) High frequency response region.
(d) Very high frequency response region.

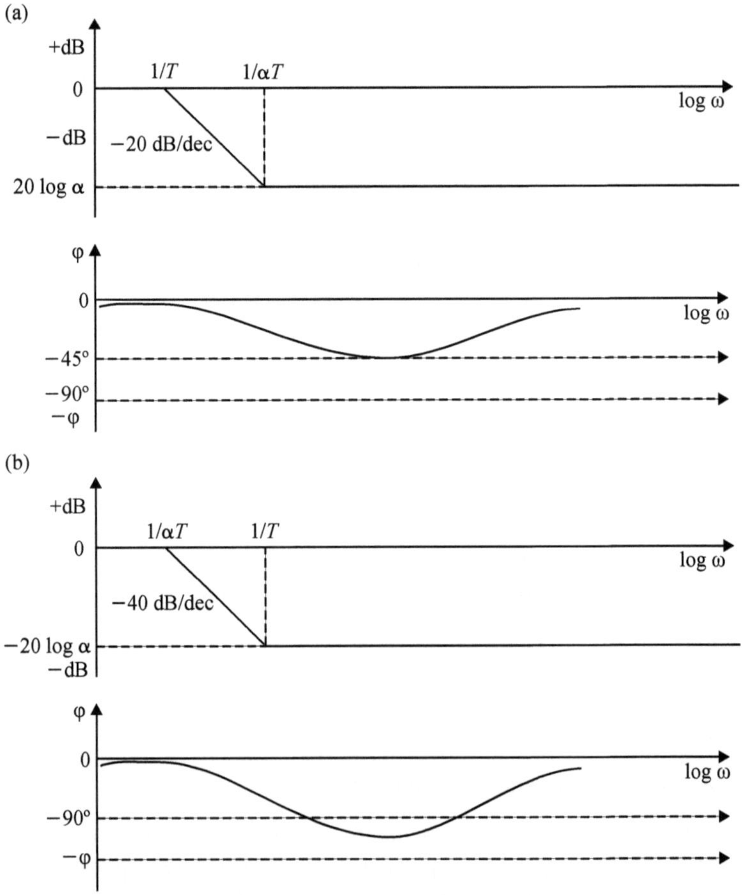

Fig. 6.5 Bode plots for a lag network

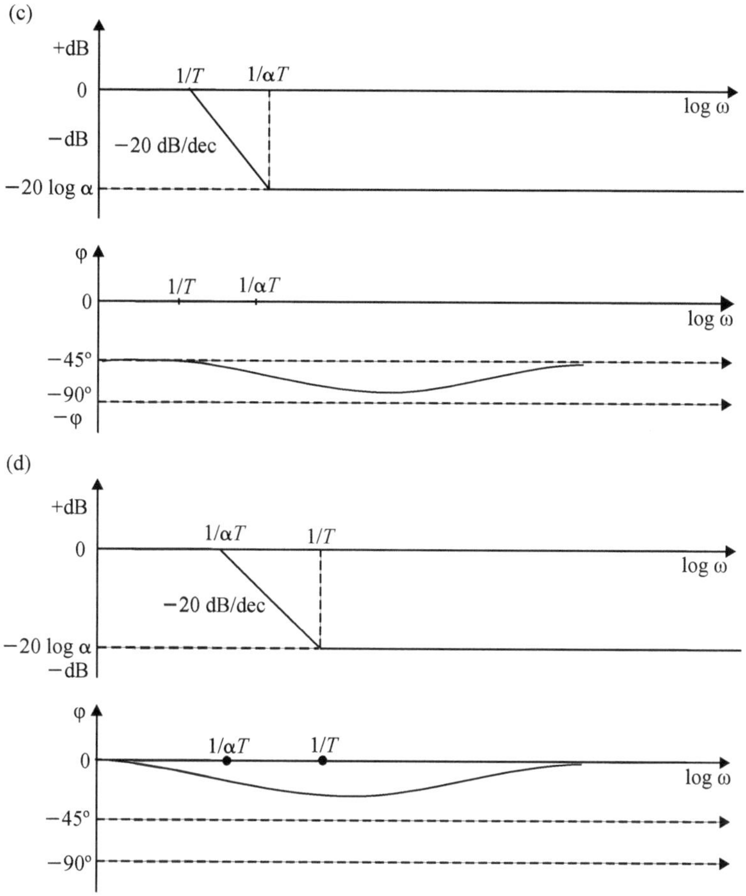

Fig. 6.5 (continued)

37. How are poles are zero of the lag network placed complex s-plane?

(a) Poles and zeros are placed at far away places on the negative real axis in the s-plane.
(b) Poles and zeros of the lag network are placed in such a way maximum phase angle lag is contributed.
(c) Poles and zeros are placed as much closer as possible.
(d) Closer to each other and nearer to the origin on the negative real axis of the s-plane.

38. How does a lag compensator perform?

(a) The excessive lead angle contributed by the system is reduced by the lag angle provided by the lag network.

(b) For a good transient response the excessive phase lag is necessary which is provided by the lag network.
(c) Since the angle contribution provided by the poles and zeros are almost negligible and provides attenuation which shifts the gain cross over frequency to the low frequency region which results in the required phase margin.
(d) Since the attenuation in the magnitude of the lag network in the middle and low frequency region is provided, the phase margin is increased.

39. What is the overall effect of lag compensator?

(a) Lag compensator improves steady state response.
(b) Lag compensator improves transient response.
(c) Lag compensator improves steady state response and does not have any influence in the transient response.
(d) Lag compensator improves transient response and does not influence the steady state response.

40. What is the effect of the lead compensator?

(a) Lead compensator improves the transient response.
(b) Lead compensator influences the steady state response.
(c) Lead compensator improves steady state response and does not influence the transient response.
(d) Lead compensator improves transient response with little influences on the steady state response.

41. What is the transfer function of a lag lead compensator?

(a)

$$G_c(s) = \frac{(1+\alpha_1 T_1 s)(1+\alpha_2 T_2 s)}{(1+T_1 s)(1+T_2 s)}.$$

(b)

$$G_c(s) = \frac{(1+T_1 s)(1+T_2 s)}{(1+\alpha_1 T_1 s)(1+\alpha_2 T_2 s)}.$$

(c)

$$G_c(s) = \frac{(1+(1/\alpha)T_1 s)(1+(1/\alpha_2)T_2 s)}{(1+T_1 s)(1+T_2 s)}.$$

(d)

$$G_c(s) = \frac{(1+(1/\alpha)T_1 s)(1+(1/\alpha_2)T_2 s)}{(1+T_1 s)(1+T_2 s)}.$$

42. For $\alpha_1 < 1$ and $\alpha_2 > 1$, which portion of the lag-lead network represents lag compensator and which portion represents the lead compensator.

(a)

$$G_{c1} = \frac{(1 + \alpha_1 T_1 s)}{(1 + T_1 s)}$$

represents lag portion.

$$G_{c2} = \frac{(1 + \alpha_2 T_2 s)}{(1 + T_2 s)}$$

represents the lead compensator.

(b)

$$G_{c1} = \frac{(1 + \alpha_2 T_2 s)}{(1 + T_2 s)}$$

represents lag network.

$$G_{c2} = \frac{(1 + \alpha_1 T_1 s)}{(1 + T_1 s)}$$

represents lag portion.

(c)

$$G_{c1} = \frac{(1 + T_1 s)}{(1 + \alpha_1 T_1)}$$

represents lag portion.

$$G_{c2} = \frac{(1 + T_2 s)}{(1 + \alpha_2 T_2 s)}$$

represents lead portion.

(d)

$$G_{c1} = \frac{(1 + \alpha_1 T_1 s)}{(1 + \alpha_1 T_2 s)}$$

represents lag portion.

$$G_{c2} = \frac{(1 + T_1 s)}{(1 + T_2 s)}$$

represents lead portion.

43. How a lag-lead network is designed?

(a) Lag portion is designed ignoring lead portion.
(b) Lead portion is designed ignoring lag portion.
(c) The lag portion poles and zeros are chosen so that they are placed closer to the origin of the s-plane and the phase contribution is negligibly small. Then design a lead compensator to meet the required specifications.
(d) None of the above.

44. What is the effect of lag compensator or bandwidth?

(a) Lag compensator increases the bandwidth.
(b) Lag compensator diminishes the band width since it is a low pass filter.
(c) At middle frequency bandwidth is increased.
(d) At frequency bandwidth is increased.

45. What is the effect of Lag-Lead compensator on bandwidth?

(a) Lead network, for a wide frequency range the bandwidth is decreased.
(b) The bandwidth is increased.
(c) At middle frequency the bandwidth is decreased.
(d) At high low frequency the bandwidth is decreased.

Answers

(1) d	(10) a	(19) a	(28) c	(37) d
(2) d	(11) c	(20) b	(29) b	(38) c
(3) a	(12) a	(21) d	(30) a	(39) c
(4) d	(13) a	(22) b	(31) b	(40) d
(5) d	(14) a	(23) d	(32) a	(41) a
(6) d	(15) a	(24) c	(33) d	(42) a
(7) d	(16) a	(25) c	(34) a	(43) c
(8) d	(17) b	(26) c	(35) b	(44) b
(9) d	(18) b	(27) d	(36) c	(45) b

Chapter 7
State Space Analysis

Multiple Choice Questions

Choose the Correct Answer

1. **The state of a system is**
   (a) the minimum initial conditions required to represent several variables in the system.
   (b) the minimum number of initial conditions that are required at any initial time together with the input so that all the information about the variables are determined in the time interval $t_0 \leq t \leq t_f$ where t_f is the finite time.
   (c) the states represent the variables which are dynamic.
   (d) the states represent output variables at different points.

2. **An autonomous system is represented by the following equation**

$$\dot{x} = Ax$$

   where x is the state variable of order n. The dimension of the A matrix is
   (a) $n \times n$ square matrix.
   (b) $n \times 1$ vector.
   (c) $1 \times n$ vector.
   (d) none of the above.

3. **Consider the following equation**

$$\dot{x} = Ax + Bu$$

© The Author(s), under exclusive license to Springer Nature Switzerland AG 2026

S. Palani, *Multiple-Choice Questions in Control Systems Engineering*,
https://doi.org/10.1007/978-3-031-93787-3_7

where x is the state vector and u is the input vector. The above equation is called as

(a) vector matrix equation.
(b) vector matrix differential equation of a continuous time system.
(c) linear algebraic equation.
(d) non-linear state equation.

4. Can non-linear system equations be represented by vector matrix equation?

(a) Yes.
(b) No.
(c) Some equations can be represented.
(d) None of the above.

5. For a linear system is it possible to find the T.F. model form state model?

(a) It is possible.
(b) It is not possible.
(c) In many cases it is not possible.
(d) None of the above.

6. For a given system the state space model is

(a) unique.
(b) non-unique.
(c) unpredictable.
(d) none of the above.

7. The eigenvalues obtained from the state equation and the characteristic equation are

(a) same.
(b) different.
(c) unpredictable.
(d) none of the above.

8. The eigenvalues from the state equations are obtained from

(a) $[\lambda I - A] = 0$.
(b) $|\lambda I - A|^{-1} = 0$.
(c) $|\lambda I + A| = 0$.
(d) $[\lambda I + A]^{-1} = 0$.

9. Consider the following vector matrix differential equations

$$\dot{x}(t) = Ax(t) + Bu(t)$$
$$y(t) = Cx(t) + Du(t)$$

The T.F. of the system is

(a)
$$\frac{Y(s)}{U(s)} = C(sI - A)^{-1}B + D.$$

(b)
$$\frac{Y(s)}{U(s)} = C(sI + A)^{-1}B + D.$$

(c)
$$\frac{Y(s)}{U(s)} = C(sI - A)B + D.$$

(d)
$$\frac{Y(s)}{U(s)} = C(sI + A)B + D.$$

10. How is state transition matrix represented in the time domain?

(a) $STM = \phi(t) = e^{At}$.
(b) $STM = \phi(t) = e^{-At}$.
(c) $STM = \phi(t) = [sI - A]^{-1}$.
(d) $STM = \phi(t) = [sI + A]^{-1}$.

11. How is state transition matrix is represented in the frequency domain?

(a) $STM = \phi(s) = [sI + A]^{-1}$.
(b) $STM = \phi(s) = [sI - A]^{-1}$.
(c) $STM = \phi(s) = [sI - A]$.
(d) $STM = \phi(s) = [sI + A]$.

12. For an autonomous system, how the transient response is obtained?

(a) $x(t) = L^{-1}[\phi(s)x(0)]$.
(b) $x(t) = L^{-1}[\phi(s) + X(s)]$.
(c) $x(t) = L^{-1}[\phi(s) - X(s)]$.
(d) $x(t) = L^{-1}[\phi(s)X(s)]$.

13. The Caley-Hamiltan theorem states that

(a) the system A matrix satisfies its own characteristic equation.
(b) the system A matrix satisfies the vector matrix differential equation.
(c) the system A matrix satisfies the algebraic equation of the system.
(d) none of the above.

14. Consider the following vector matrix equation with the initial state $x(0)$

$$\dot{x}(t) = Ax(t) + Bu(t)$$

The solution for this is

(a)

$$x(t) = e^{At}x(0) + \int_0^t e^{A(t-\tau)}Bu(\tau)d\tau.$$

(b)

$$x(t) = e^{At}x(0) + \int_0^t e^{-A(t-\tau)}x(\tau)Bu(\tau)d\tau.$$

(c)

$$x(t) = e^{-At}x(0) + \int_0^t e^{A(t-\tau)}Bu(\tau)d\tau.$$

(d)

$$x(t) = e^{At}x(0) + \int_0^t e^{-A(t-\tau)}Bu(\tau)d\tau.$$

15. A linear time invariant continuous time system is said to be state controllable iff

(a) the output is the desired one for the given input.
(b) the system is the desired one for the given input.
(c) it is possible to transfer any initial state $x(t_0)$ to any final state $x(t_f)$ in the time interval $t_0 \leq t \leq t_f$ without putting any control in the choice of the input.
(d) for any input, the out is bounded.

16. Consider the following state equation

$$\dot{x}(t) = Ax(t) + Bu(t)$$

The system is said to be controllable iff the

(a) composite square matrix

$$Q = [B \;\; AB \;\; A^2B \ldots A^{n-1}B]$$

has full tank.
(b) if the T.F. of the system has no pole zero cancellation.
(c) if the rank of the controllability matrix Q is not full.
(d) any one of the above except (c).

17. For a continuous time system the system is output controllable if the controllability composite matrix

(a)

$$Q = [CB \;\; CAB \;\; CA^2B \ldots CA^{n-1}B]$$

has full rank.

(b) there is no pole zero cancellation in the system.
(c) any one of the above.
(d) none of the above.

18. Consider the following system

$$\dot{x} = \begin{bmatrix} -1 & 0 \\ 0 & -2 \end{bmatrix} x + \begin{bmatrix} 1 \\ 1 \end{bmatrix}$$
$$y = [1 \quad 1]x$$

Find the state and output controllability,

(a) state controllable output not controllable.
(b) output controllable but not state controllable.
(c) both state and output uncontrollable.
(d) both state and output controllable.

19. Considerable system described by the following equation

$$\dot{x}(t) = \begin{bmatrix} -1 & 0 \\ 0 & -1 \end{bmatrix} x + \begin{bmatrix} 0 \\ 1 \end{bmatrix} u$$

Is there any pole zero cancellation in the system and what is the status of controllability? What is the rank of the matrix?

(a) There is no pole zero cancellation, hence the system is controllable.
(b) The rank is 2 and the system is controllable.
(c) The system rank is one and the system is uncontrollable.
(d) Pole zero cancellation takes place and hence the system is uncontrollable.

20. A linear time invariant continuous is said to be observable

(a) if the output is measurable.
(b) if every initial state $x(t_0)$ can be determined by measured the output $y(t)$ in a finite time interval $t \leq t \leq t_f$.
(c) if every initial state $x(t_0)$ is measuring by adjusting the output $u(t)$ in a finite time interval $t \leq t \leq t_f$.
(d) none of the above.

21. For the continuous time system described below, what is the observability condition

$$\dot{x}(t) = Ax(t) + Bu(t)$$
$$y(t) = Cx(t)$$

The observability condition is

(a) the composite square matrix

$$S = [A^T \;\; A^T C^T \;\; A^{T^2} C^T \ldots (A^T)^{n-1} C^T]$$

should have full rank.

(b) the composite square matrix

$$S = [C^T \;\; A^T C^T \;\; A^{T^2} C^T \ldots (A^T)^{n-1} C^T]$$

should have full rank.

(c) there should be pole zero cancellation in the T.F.

(d) none of the above.

Answers

(1) b	(4) a	(7) a	(10) a	(13) a	(16) d	(19) c
(2) a	(5) a	(8) b	(11) b	(14) a	(17) c	(20) b
(3) b	(6) b	(9) a	(12) a	(15) c	(18) d	(21) b

MIX
Papier aus verantwortungsvollen Quellen
Paper from responsible sources
FSC® C105338

If you have any concerns about our products,
you can contact us on
ProductSafety@springernature.com

In case Publisher is established outside the EU,
the EU authorized representative is:
Springer Nature Customer Service Center GmbH
Europaplatz 3, 69115 Heidelberg, Germany

Printed by Libri Plureos GmbH
in Hamburg, Germany